DATE DUE

BRODART, CO.   Cat. No. 23-221

Genentech

synthesis

*A series in the history of chemistry, broadly
construed, edited by Angela N. H. Creager, John E.
Lesch, Stuart W. Leslie, Lawrence M. Principe,
Alan Rocke, E. C. Spary, and Audra J. Wolfe, in
partnership with the Chemical Heritage Foundation*

Genentech

The Beginnings of Biotech

SALLY SMITH HUGHES

The University of Chicago Press
Chicago and London

SALLY SMITH HUGHES is a historian of science at the
Bancroft Library at the University of California,
Berkeley. She is the author of *The Virus: A History of the
Concept* and the creator of an extensive collection of
in-depth oral histories on bioscience, biomedicine, and
biotechnology.

The University of Chicago Press, Chicago 60637
The University of Chicago Press, Ltd., London
© 2011 by The University of Chicago
All rights reserved. Published 2011.
Printed in the United States of America

20 19 18 17 16 15 14 13 12 11   1 2 3 4 5

ISBN-13: 978-0-226-35918-2 (cloth)
ISBN-10: 0-226-35918-2 (cloth)

Library of Congress Cataloging-in-Publication Data

Hughes, Sally Smith, author.
    Genentech: the beginnings of biotech /
Sally Smith Hughes.
        p. cm. — (Synthesis)
    ISBN-13: 978-0-226-35918-2 (cloth: alkaline paper)
    ISBN-10: 0-226-35918-2 (cloth: alkaline paper)
1. Genentech, Inc.—History. 2. Biotechnology
industries—United States—History. 3. Biotechnology—History. I. Title. II. Series: Synthesis
(University of Chicago Press)
    HD9999.B444H85 2011
    338.7′66060973—dc22   2011004666

♾ This paper meets the requirements of
ANSI/NISO Z39.48-1992 (Permanence of Paper).

Dedicated to the memory of
Janet Wentworth Smith (1910–2007)
and to my children,
Dylan, Amy, and Casey

Contents

Prologue

Genentech: The Beginnings of Biotech is the story of a pioneering genetic-engineering company that inspired a new industrial sector, transforming the biomedical and commercial landscapes ever after. Yet the enormous success of Genentech—today an icon of a worldwide biotechnology industry—appears to confirm an industrial myth: that the company's achievements and the industry it fostered were straightforward, preordained, and inevitable events. This interpretation is far from the mark.

Genentech was by almost any measure an inauspicious and improbable enterprise. The firm arose in the spring of 1976 as the unlikely vision of two naive entrepreneurs: Herbert Boyer, a professor of microbiology at the University of California, San Francisco; and Robert Swanson, an unemployed venture capitalist. Its name, a contraction of *genetic engineering technology*, captured its extraordinary agenda: to apply the radically new technology of recombinant DNA in engineering bacteria to make insulin, growth hormone, and other important pharmaceuticals. But no one had ever employed the technology as an industrial process, much less built a business upon it and tried to make a profit from it. Could the company produce, before the money ran out, the novel therapeutic substances Boyer and Swanson had in mind? Prominent molecular biologists thought not.

The making of Genentech was in fact racked by problems, internal and external. The science did not always work. Swanson and Boyer struggled to raise and sustain venture-capital backing and to forge contracts with a skeptical pharmaceutical industry. Scientists, managers, and attorneys needed cajoling and incentives to abandon university and corporate positions to join a company broadcasting risk at every level. In a tumultuous period of science politics, pending government regulation and adverse public opinion threatened substantial interference. Swanson and Boyer had to devise corporate structure and erect intellectual property protection, giving Genentech a chance to compete against far larger and well-established companies with vastly deeper pockets. He needed to somehow balance a freewheeling, university-like culture against the dire need to eke out a profit, file for patents, and above all make marketable products. The cards appeared stacked against the venture's success. External conditions likewise menaced Genentech's agenda to put genes to work. The firm struggled to get off the ground during a time of soaring public apprehension over biology's new power to engineer bacteria of possible threat to human health and safety. As if the deepening scientific, political, and cultural ferment were not enough, the infant company had to also navigate federal guidelines for recombinant DNA research, face the threat of restrictive legislation, and run the gauntlet of legal unknowns in patenting living things.

Genentech's future rested on technological innovation, business acumen, human dedication, and a freewheeling, can-do culture strikingly different from anything the pharmaceutical industry offered. Its handful of irreverent scientists captured a swiftly expanding audience as they cloned and expressed three medically significant genes in three successive years: human insulin, human growth hormone, and human interferon—the latter a supposed miracle drug predicted to cure cancer and other ills. Genentech's come-from-behind scientific contests against prestigious university teams led to biotechnology's first research and development contracts with pharmaceutical companies. By 1980 the firm was setting scientific and business standards that a handful of fledgling biotech companies and academic entrepreneurs would attempt to emulate and adapt to their specific needs. Molecular biology was becoming practical, profitable, and controversial in a manner never before experienced.

Remarkably, just four years after its creation, Genentech became the overnight darling of Wall Street. Its public stock offering raised over $38 million, as share price rocketed from $35 to $89 in a wild first few minutes of trading. It was the largest gain in stock market history, making headlines around the world. To the investment and business communities and a riveted public, agog at what they were witnessing, the company confirmed that genetic engineering could build a business, attract major money, and promise lifesaving pharmaceuticals as well. Genentech's spectacular success launched a period of speculative frenzy over biotechnology as a revolutionary approach for creating novel products, generating incalculable profits, and fashioning a new industrial sector.

This intimate and people-centered history traces the seminal early years of a company that devised new models for biomedical research, business, and culture, in the process introducing a novel creation—the entrepreneurial biologist. As a prime instigator of a mounting commercialism blossoming in biology of the 1970s, Genentech in large measure recast the aspirations, direction, and culture of life science and set the stage for the formation of a biotechnology industry.

Acknowledgments

The long journey to a published book began in 1992 with a modest project at the University of California, San Francisco, to interview a few of its movers and shakers in molecular biology and biotechnology. Significantly, Herbert Boyer and William Rutter were among them. Little did I then know that this was not just another fascinating project in my many years of interviewing in and writing on recent science and technology. The interviews with Boyer and Rutter brought to life important themes: the contingent process of making a basic science practical, patentable, and profitable; the impact of political controversy and monumental legal questions; and the historical significance of captivating personalities, sometimes at odds with one another. Whether I fully realized it or not, seeds for a historical work had begun to germinate.

When I returned in the late 1990s to the Bancroft Library at the University of California, Berkeley, Daniel E. Koshland Jr.—a prominent Berkeley biochemistry professor, Lasker Award winner, and former chief editor of the journal *Science*—made a generous donation to inaugurate an oral history and archival project documenting bioscience and biotechnology in the San Francisco Bay Area. I am inordinately grateful to him for his support and friendship. To my sorrow, Dan died in 2007 without seeing this book, a tangible product of his benevolence.

In 2001 the baton was even more generously taken up by the Genentech Foundation in sponsoring an oral history and archival program on its foundational period. The rich trove of original oral and written documentation assembled over the next few years became a central source for the present book. Bob Swanson, Genentech's cofounder and first CEO, permitted me access to documents in his private possession and in corporate archives at Genentech—documents that had never before been available to outsiders. The history recounted in these pages is far richer as a result. It was with great sadness that I learned of Bob's premature demise of a brain tumor in 1999. A small comfort is the eight hours of interviews on Genentech's early years that he recorded with me in 1996 and 1997.

Aside from the colorful personalities and working lives of Genentech's first generation—a carrot for an author fascinated by human agency—participants' enthralling accounts of erecting a novel company upon controversial basic science were an immediate draw. As I moved along in the oral history project, even more so were the insights I gathered regarding the wider meanings of Genentech's early history for science, industry, and society. I am deeply grateful for the support and especially for the free rein the company gave me to conduct research and interviews as I saw fit. The company's sole condition was to submit draft interview transcripts to its legal department for review exclusively for legal matters. The attorneys never once responded; the oral histories appear in final form as the Bancroft Library alone produced them. The entire interview series on Genentech and on science and technology more generally is available in searchable format at http://bancroft.berkeley.edu/ROHO/projects/biosci/. One incalculable benefit of the oral history process was the personal relationships created that allowed me later in the course of writing to return time and again to Genentech folks with questions and requests. The history told herein is fuller and more accurate as a result.

A fellowship at the National Humanities Center in 2006–7 gave me a luxurious nine months to focus on writing. I thank the National Endowment for the Humanities for supporting my fellowship. The center is a blessing to scholars in the uninterrupted time and support it provides fellows for concentrated writing and daily opportunities for conversations with scholars in the full sweep of the humanities. I am particularly indebted to participants in the history writing group for the congenial atmosphere they provided for critiques of early draft chapters. This book has also greatly benefited from my various presentations at Berkeley, Duke University, the History of Science Society, Society for the History of Technology, and elsewhere, as well as from comments by two anonymous readers selected by the University of Chicago Press.

My deepest thanks to the following individuals for critiquing draft chapters: Henry Bourne, Herb Boyer, Stan Cohen, Bob Cook-Deegan,

Roberto Crea, Martha Dill, Jane Gitschier, Dave Goeddel, Ellie Goldberg, Herb Heyneker, Tom Kiley, Dennis Kleid, John Lesch, Art Riggs, Bill Rutter, Milt Schlesinger, Sondra Schlesinger, Winni Sullivan, Eric Vettel, Ron Wetzel, and Dan Yansura. Special thanks also to editors Karen Darling and Audra Wolfe, who helped to shape and shepherd this book.

1

Inventing Recombinant DNA Technology

I looked at the first gels [in the first recombinant DNA cloning experiment], and I can remember tears coming into my eyes, it was so nice. I mean, there it was. You could visualize your results in physical terms, and after that we knew we could do a lot of things.
    Herbert W. Boyer, March 28, 1994[1]

Modern biotechnology originated in 1973 with the invention of recombinant DNA technology, a now-universal form of genetic engineering. It entails recombining (joining) pieces of DNA in a test tube, cloning (creating identical copies of DNA) in a bacterium or other organism, and expressing the DNA code as a protein or RNA molecule. It soon vastly extended the power and scope of molecular biology, penetrated several industrial sectors, and became a cornerstone of a new industry of biotechnology. Yet technological power and potential cannot alone explain its first commercial application—at the biotechnology company Genentech in the mid-1970s. Stanley Cohen and Herbert Boyer, the two inventors, had designed the technique for basic-science research. But they immediately foresaw its practical applications in making plentiful quantities of insulin, growth hormone, and other useful substances in bacteria. Despite their common starting point, Cohen and Boyer chose different avenues for industrializing recombinant DNA technology. Why they did so was a matter of personality and professional commitments. It was also a matter of the national environment in the U.S. of the 1970s—a pivotal decade of raging debate in science politics, major dilemmas and decisions in constitutional and patent law, and cultural, attitudinal, and personal challenges as commercial interests first entered molecular biology full force.

TWO SCIENTISTS ON CONVERGING PATHS

Herbert Wayne Boyer was born in 1936 into a blue-collar family and grew up in the little town of Derry, thirty miles from Pittsburgh in the coal-mining country of western Pennsylvania. His father had left school in eighth grade and eventually found work as a railroad brakeman and conductor. His mother married straight out of high school and stayed home to look after Herb and a younger sister. Herb earned pocket money by mowing lawns, delivering newspapers, and doing other odd jobs of a middle-American boyhood. He hunted and fished with his father and developed an abiding love of the outdoors. All four Boyers played at least one musical instrument and regularly got together with family and friends to play country-western music—bright spots in an otherwise workaday world. Herb's first years at Derry Borough High School were a steady round of football, basketball, baseball, and girls—anything but academic achievement. He was on "a rather perilous course of delin-quency,"[2] he later admitted. It took a no-nonsense football coach and teacher to jolt Herb out of his apathy. "Pat Bucci straightened me out," he subsequently observed. He began belatedly to focus on schoolwork. Coming into his own, he was elected junior- and senior-class president and voted most athletic. But the limited vistas of a small railroad town felt more and more confining. One way or another, he had to get out. Herb resolved to go on to college, the first in his family to do so. He was off to troll wider horizons but destined never to lose the down-to-earth practi-cality and lack of pretension of his blue-collar upbringing.

Stanley Norman Cohen is also the first child and only son of parents whose formal education ended with high school. His father was a small businessman who tried his hand, never very successfully, at several trades in and around their home in Perth Amboy, New Jersey, a town just southwest of New York City. Stan's mother worked for a time as a secretary to make ends meet. Stan was born in 1935 and raised as an only child un-til the birth of a sister when Stan was almost ten. Tight finances, gentle discipline, and parental ambition for their children to rise in the world largely defined home life. While Boyer needed Coach Bucci's interven-tion to provoke his attention to schoolwork, learning came naturally and

Fig. 1. Herb Boyer (*second row on left*) in his high school laboratory, in Derry, Pennsylvania, in the mid-1950s. (Photographer unknown; photograph courtesy of Herbert W. Boyer; credit: Cover Studio, 504 Main Street, Johnstown, PA.)

at an early age for Cohen. No adult had to build discipline in young Stan. "I suppose," he recalled, "that overall I wasn't much of a wayward kid, so there really wasn't a lot of need for discipline."[3] He and his father, a frustrated inventor, spent off-hours in the basement doing small wiring and mechanical projects. Cohen credits his father with sparking his interest in how things work—sparking his interest in *science*. From the start he was motivated to achieve, and achieve he consistently did. In high school he was editor of the school paper and associate editor of the yearbook. By then his scientific interests centered on biology, which to Stan meant becoming a physician. He now had a goal that would move him beyond the narrow scope of his upbringing. Yet he would remain stamped with the work ethic, professional ambition, and respect for knowledge of his Jewish heritage.

Boyer and Cohen, with only slightly more than a year between them, came of age in the early 1950s. Both were financially strapped; both could expect no financial assistance from their families; both chose colleges close to home. In 1954 Boyer entered Saint Vincent College, a liberal-arts institution run by Benedictine monks in Latrobe, Pennsylvania, a few easy miles from Derry. He lived at home to save money and hitchhiked or rode the bus to and from classes. His father, a railroad man, refused to learn to drive, let alone to buy a family car. Boyer majored in biology and chemistry, intending to go on to medical school. A chance class assignment suggested another direction. More than five decades later, Boyer recalled the shift with the clarity of a formative moment:

> We had a brand-new, shiny [cell physiology] textbook with a blue and white cover. Each of us was assigned a chapter, and we had to give a seminar on it. Which one did I get? "The Structure of DNA." This was 1957, and the buzz of DNA was just getting into the textbooks. . . . I was really taken with the Watson-Crick structure of DNA and this started my fascination with the heuristic value of the structure.[4]

A sign of his new infatuation was Boyer's Siamese cats, Watson and Crick. In 1958 Saint Vincent awarded Boyer a bachelor's degree in biology and chemistry.

Boyer applied and failed to enter medical school, a D in metaphysics being his nemesis. He settled on graduate school at the University of Pittsburgh, partly to improve his grades and reapply to medical school, partly because "a small-town boy doesn't stray too far from home."[5] He craved intellectual stimulus and found it in the heady research of a bacterial genetics laboratory at a watershed moment in molecular biology. Watson and Crick's discovery of 1953 had launched an avalanche of work on major questions—prime among them, the nature of the genetic code and the mechanism of protein synthesis. Much of this research transpired in bacteria, employed by experimentalists for their relative simplicity as compared to the animal kingdom. Boyer thrived on the lab's scientific ferment and freeform discussion on genetic exchange and recombination in bacteria. "That [lab]," Boyer recalled, "was my [scientific] awakening."[6]

In 1959, at the end of his first year of graduate school, Boyer married his high school sweetheart, Marigrace Hensler, a biologist in her own right. She gamely supported the couple, as Boyer tackled a near-intractable experiment on deciphering the genetic code. Breaking the code was the foremost problem in molecular biology of the day, one that only a supremely ambitious—or naive—graduate student would agree to take on. Boyer did and plugged away, even after two biochemists broke the code in 1961. "Boyer," a future colleague commented, "consistently tried big things without knowing whether they could or should work."[7] He managed to squeeze out enough data to complete a dissertation. His attraction to challenging problems would become a mark of his professional career. Boyer was setting an enduring pattern. Below the casual surface lay ambition and tenacity. In 1963 Boyer earned a doctoral degree in bacteriology.

Cohen chose Rutgers University, a few short miles from Perth Amboy. Rutgers offered the most scholarship support and was close to his ailing father. Studious as ever, Cohen worked hard but carried to extremes his resolve to make a life beyond academics. He joined the university debating team, took up the guitar, and tried his hand at writing pop songs, one of which reached the hit parade. This flurry of extracurricular activities, predictive in its intensity, failed to dent his academic performance. In 1956 he graduated magna cum laude from Rutgers. That fall Cohen entered medical school at the University of Pennsylvania, a major draw being the substantial scholarship funds it provided. His first taste of basic research in the second year led to a summer research position in London and to the publication of his first paper. He took time off that summer to wander the cafés of Europe, supporting himself by singing and strumming the banjo. "It was a wonderful time,"[8] he recalled, remembering the freedom and lack of responsibility. Life from then on would never again be as carefree, but banjo and song would remain outlets for life. In 1960 Cohen graduated from Penn with a degree in medicine. Within a whirlwind five-year period, swinging from the East Coast to the South, Cohen completed an internship, a two-year research position at the National Institutes of Health (NIH), and a residency in medicine. In 1961 he married Joanna Wolter, and they eventually had two children.

Boyer's career took a less peripatetic course. He went straight from Pittsburgh to Yale as a postdoctoral fellow in microbiology. There he joined a lab focused on genetic exchange and recombination in bacteria. He became fascinated with the restriction enzymes of bacteria— enzymes that cut up and destroy foreign DNA entering the bacterial cell.[9] The word just emerging in the 1960s was that certain types of restriction enzymes sever DNA at unique sequences in the molecule. Perhaps, Boyer and others recognized, one could use these strange enzymes to clip DNA into well-defined fragments and map its structure. He suspected early on that restriction enzymes were going to be "very helpful enzymes" for the precision cutting, recombination, and characterization of DNA.[10] The suspicion was prophetic: his career-long passion would become restriction enzyme research and genetic manipulation. Boyer now lived and breathed his science. After a night on the town, he would return to

Fig. 2. Stan Cohen, circa 1956, yearbook photograph, Rutgers University. (Photographer unknown; photograph courtesy of Stanley N. Cohen.)

the lab or rise in the dark to observe an experiment. But the folks at home were stymied. "What are you doing?" his father would ask. "Restriction endonuclease modification," he would glibly answer, using the technical term for his research area. He would then pause for his father's inevitable retort, "Well, what good is it? What are you going to do with that?" Boyer would respond, "I don't know—cure the common cold."[11] His answer was dismissive, but his father's question prompted him to ponder the practical utility of his research.

Cohen meanwhile had begun a postdoctoral research fellowship (1965–67) in molecular biology at Albert Einstein College of Medicine in New York. It was here he stopped wavering between a career in medicine or science. He decided to pursue both, apparently expecting the rewards of a dual career to outbalance its tensions and frantic pace. He took up research on plasmids, tiny rings of DNA in the cytoplasm of bacterial cells that reproduce outside the main chromosome. Plasmids typically carry antibiotic resistance genes that can pass from one bacterium to another, spreading the resistance problem. The study of plasmids was at the time a quiet backwater and to Cohen consequently appealing. Scientists interested in genetic exchange and gene regulation mainly studied viruses, which had been a central focus of molecular studies from the 1930s on.[12] Cohen reasoned that his heavy clinical responsibilities would make successful competition with "hotshot" molecular biology labs difficult if not impossible. Plasmid research seemed a perfect fit: he knew the necessary molecular and biochemical techniques, and the growing medical problem of antibiotic resistance was an appropriate topic for a physician.[13] He was correct in every regard except for expecting the field to remain tranquil. It was about to explode, and Cohen would find himself at its epicenter.[14]

By 1968 Cohen was intent on finding a faculty position. One of his mentors had collegial associations with several members of the biochemistry department at Stanford University. The connections led to a job offer, but not in biochemistry. Some years earlier, the clinical departments at Stanford Hospital in San Francisco had moved south to join the preclinical departments on Stanford's Palo Alto campus—a reorganization aimed at bringing basic science and clinical medicine into geographic and intellectual proximity. Recognizing Cohen as one of a new breed of physician-scientists the school sought to attract, the Department of Medicine offered Cohen an assistant professorship in its Division of Hematology. Cohen, drawn by the California climate and lifestyle, accepted and in 1968 moved with his wife to the sun-swept campus in Palo Alto.

He was disheartened to find that no one in the department shared his fascination with molecular genetics. He turned for advice to Arthur Kornberg, the powerful chairman of Stanford's biochemistry department. By virtue of his Nobel Prize, academic position, and forceful personality, Kornberg was a figure to reckon with. Not one to mince words, he told Cohen that plasmid research was an uninteresting line of investigation. The irony of the remark would soon become apparent. "So this wasn't a very comforting introduction to Stanford,"[15] Cohen recalled. His understatement skated over what must have been an unsettling blow: he had come to Stanford considering that shared scientific interests and collegial ties might lead to a joint appointment in the Department of Biochemistry. Instead, Kornberg made it clear that Cohen's association with the department was at best to be informal. Kornberg almost never granted joint appointments and also believed that only a rare individual could optimally perform both clinical medicine and basic research.[16] Despite the tepid welcome and even after his own lab was operating, Cohen, according to biochemist Paul Berg, "hung around in [the Department of] Biochemistry most of the time."[17]

Berg exaggerated, but Cohen indeed thrived on the department's stimulating intellectual exchange and had access to its electron microscope and other equipment lacking in his home department. He regularly attended biochemistry seminars and benefited from the chance "to bounce ideas off people in that department."[18] He particularly profited from discussing ongoing departmental research on DNA ligation (joining) and DNA uptake by animal cells. In turn, he shared with biochemistry colleagues his work on plasmid isolation and characterization. But the research Cohen published in the early 1970s was not done in collaboration with Stanford biochemistry faculty. As he stated forcefully in 2010: "Notwithstanding Kornberg's notion that all important scientific knowledge at Stanford originated in the Department of Biochemistry, the work on DNA replication in that department had absolutely zero impact on my research. Similarly, the work on biochemical methods of dAT joining"—a biochemical method for joining DNA fragments—"by Berg and others did not impact my work."[19]

Cohen was using plasmids to transport genes and DNA fragments into bacteria; Berg's group and others in Stanford biochemistry were using viruses as transport vehicles. The department was at the forefront of techniques for joining DNA molecules from different sources. In 1972 Berg and his lab succeeded in making the first recombinant DNA molecules in a test tube.[20] However, no member of the Stanford biochemistry faculty or anyone elsewhere created a method for cloning DNA. The various biochemical approaches Berg and the others developed for recombining DNA were technically complicated and required a battery of enzymes and skills "probably beyond the ability of most labs,"[21] as Berg himself admitted. The dire need for a simple and efficient method for joining and replicating genetic material continued to go unanswered.

Boyer was also looking for a faculty appointment as his postdoctoral years wound down. Hearing of an open position in the microbiology department at the University of California, San Francisco, he applied, lured less by the medical school's reputation—mediocre—than by the charms of the Golden State. He loved cowboy and Indian movies and had always wanted to visit California. Boyer accepted an appointment as assistant professor at an annual salary of $12,500, and in 1966 moved with his family to join the basic-science faculty. Denied the promised lab space in one of the new research towers, he settled, much annoyed, into cramped laboratories in a department chaired by an old-school microbiologist with no interest in molecular genetics. Boyer found diversion in the rich protest and counterculture movements of the Bay Area in the 1960s, later telling a reporter that he had participated in nearly every antiwar rally in the vicinity.[22] Like many of his colleagues, he was captivated by the growing technical power of the new genetics to manipulate the stuff of life. But he was also troubled by its social and ethical implications.[23] How far should scientists go in their ability to "tamper with life"? In one of many occasions, he and others at a molecular biology meeting spent an entire evening discussing how genetic engineering might affect society for good and ill.[24] It was not the last time Boyer would participate in gatherings on social accountability in science, although as the tables turned, he would find himself protesting governmental efforts to restrain science. His social activism served as a diversion at a time when his research, despite long, laborious hours, was only marginally productive. He spent four years studying a restriction enzyme that he eventually concluded cut the DNA molecule in an unhelpfully random fashion. What drove him was to find one that made a unique and predictable break.[25] Discouraged by his lack of research progress and feeling like a fish out of water in his department, he thought of searching for a position elsewhere. Then circumstances began to improve.

UCSF administrators, determined for more than a decade to turn a second-tier medical school into a premier research institution, had by the late 1960s decided that the Department of Biochemistry should lead the way. An individual widely credited with facilitating the revitalization arrived in 1968 in the driving and dapper person of William J. Rutter.[26] The tirelessly resolute biochemist, a strategic thinker if there ever was one, appeared under the ambitious banner of building a biochemistry department that would apply breaking knowledge in molecular biology and biochemistry to research on the complex genetic mechanisms of higher organisms. It was a deliberate departure (in which he was not alone) from molecular biology's traditional focus on bacteria and viruses and an agenda requiring the latest techniques to decipher the complicated genomes of higher organisms.[27] Rutter's multidisciplinary research strategy and the cooperative interdepartmental culture he sought to foster were highly compatible with Boyer's research interests and collaborative scientific style. After the arrival of biochemist Howard Goodman in 1970, one of Rutter's early recruits, Boyer spent ever more time in the biochemistry department, working closely with Goodman on restriction enzyme problems. It became a second academic home—for seminars, chitchat, and the collegiality and conviviality he thrived on. The biochemistry department, Boyer recalled, "got to be a very exciting place in the early seventies for me."[28] Like Cohen, he had found an environment far more compatible with his research interests than his own department.

Cohen meanwhile juggled a staggering workload in three different areas. He had demanding clinical and basic research obligations and was also collaborating and publishing on a computer-based system for identifying drug interactions.[29] Yet somehow his productivity only increased. In 1970–72 he published thirteen papers, including nine on plasmids. By mid-1972 he and two assistants had developed a system for removing plasmid DNA from bacterial cells, shearing it into pieces in a blender, and inserting single molecules of plasmid DNA into bacteria to study plasmid structure and antibiotic resistance. But the procedure was slow and inefficient. The shearing process broke the plasmid DNA into a welter of random-length fragments, making selection and study difficult, and only on rare occasion did the plasmid DNA enter the bacterial cells.[30] Cohen pondered these deficiencies of his plasmid transfer system as he began to organize a conference on plasmid research scheduled for November 1972 in Honolulu.

Unknown to Cohen at the time, the Boyer lab that year had made a related discovery. A graduate student had isolated a restriction enzyme (the soon to be widely employed EcoRI) that cut DNA predictably at a specific position in the molecule—exactly the characteristics that Boyer had long sought. Especially exciting was the finding—by scientists in Stanford biochemistry and genetics, to whom Boyer had donated the enzyme in generous quantity—that EcoRI did not cut evenly through the double strands of the DNA molecule.[31] Instead, it made a staggered cut, creating two single projecting DNA strands. Each single strand was able to bond with a complementary DNA strand, as one Velcro strip unites with another. The Goodman and Boyer labs confirmed the Stanford finding by sequencing the DNA site cut by EcoRI and determining the order of the nucleotide subunits composing the restriction site.[32] The idea of using cohesive or "sticky ends," as they were called, to join DNA fragments had been around for a decade or so. In fact, several groups in Stanford biochemistry were chemically synthesizing artificial sticky ends, attaching them to DNA fragments, and using the sticky ends to splice one fragment to another.[33] The procedures were long and tedious. Boyer's enzyme with its natural ability to create sticky ends in just one step offered a substantial leap in ease and efficiency in splicing together DNA pieces to form recombinant molecules. Trying to capitalize on these features, Boyer and Robert Helling, a biochemist on sabbatical leave in Boyer's lab, were using the EcoRI enzyme in attempts to combine DNA fragments. It was the summer of 1972 and they were getting nowhere.

THE COLLABORATION

Cohen, meanwhile, was organizing the Honolulu plasmid conference and belatedly got word of Boyer's as-yet-unpublished work on the new restriction enzyme. Seeing its possible relevance to characterizing plasmid DNA, he issued Boyer, whom he had never met, a last-minute invitation to attend the conference.[34] Boyer recognized a golden chance to talk about EcoRI and agreed to come. In November Cohen and Boyer arrived in Honolulu for the conference, neither knowing the details of the other's research. As it came time for Boyer to present, Cohen listened raptly to his description of EcoRI's properties. His mind lit up when he heard that the enzyme cut DNA molecules predictably and reproducibly into unique fragments with sticky ends. In a flash of insight, he wondered: could one use Boyer's enzyme to sever a plasmid precisely and use the sticky ends to attach a second DNA fragment? The hybrid plasmid might then be inserted into bacteria for cloning.[35] The startling concept, if found to work, might solve the randomness and inefficiency of his plasmid transfer procedure. He urgently needed to talk with Boyer.

The occasion arose after a long day of presentations in a stuffy conference hall. Cohen and Boyer decided to take a stroll with some colleagues, eager to stretch their legs and take in the balmy air of a Hawaiian evening.[36] The walk gave Cohen and Boyer a chance to talk about the ongoing experiments in their labs. In a flash it struck them that they might have between them the makings of a method for joining and cloning DNA molecules. Pausing at a delicatessen near the beach at Waikiki, the group settled into a booth and ordered sandwiches and beer. Cohen and Boyer grew increasingly "jazzed," as Boyer later put it, about the potential synergism of their separate approaches for isolating and copying selected DNA fragments.[37] But would the ideas brainstormed over beer and deli sandwiches pan out in actual experiment? Cohen proposed a collaboration to find out. Boyer's first impulse was to donate some of his enzyme, as he had done for the Stanford scientists, and let Cohen conduct the experiment on his own. Cohen recalled saying, "Well, that doesn't seem quite fair. Your lab has spent a lot of time isolating the enzyme and we should do this as a collaboration."[38] Also to the point, Cohen needed the Boyer lab's expertise in restriction enzymes for the experiment to transpire as conceived. Boyer agreed to collaborate.

Fig. 3. Cartoon by Dick Adair re-creating the Honolulu delicatessen conversation in which Cohen and Boyer agreed to collaborate, November 1972. A figure strikingly like Cohen is depicted at right, a Boyer likeness is absent. (Reproduction of cartoon by kind permission of Dick Adair.)

Shared interests and the need for combined expertise and resources brought together two very different personalities. In many ways, they were polar opposites—in manner, demeanor, and approach to life. Boyer came across as gregarious, relaxed, and unassuming. Open to new ideas, he was willing to gamble on possibilities. Cohen struck others as private, circumspect, and exacting of himself and colleagues. Both men were inveterate workaholics and passionate about their science. But their passion manifested in contrasting manners. Boyer ran an expansively chaotic lab and preferred brainstorming over a beer. Cohen headed a small, self-contained lab group and often discussed research in the quiet of his office. Even in appearance, the contrast was striking. Boyer sported a mop of unruly brown hair, an open and cherubic face, a robust figure, and attire of jeans, running shoes, and leather vest that stretched the limits of casual. A reporter later described him as "a baroque angel in blue jeans."[39] Cohen was trim, bearded, balding, and bespectacled. He dressed casually but neatly in slacks and sweater or sport jacket. He was the quintessential image of the professor, solid citizen, and serious intellectual.

In January 1973 Cohen and Boyer began the experiment outlined in Honolulu, working it into their ongoing projects and initially giving it no exceptional priority. The specific knowledge and technical expertise of each lab defined a natural division of labor: the Cohen laboratory handled the plasmid isolation and transfer work, Boyer's the enzymology and electrophoresis studies. In a stroke of fortune, Boyer had visited colleagues on his return from Honolulu who had demonstrated their method for staining DNA fragments with a fluorescent dye that made the fragments stand out vividly on electrophoresis gels.[40] He brought the technique back to Bob Helling, still pursuing sabbatical research in Boyer's lab, who did further work correlating fragment size and mobility on the gels.[41] But the ultimate outcome of the experiment was anything but clear. As Cohen later observed, "There are some people who think that once a method of biochemical joining DNA ends was worked out, it was obvious that the chimeric [recombinant] DNA could be cloned. That's easy to say in retrospect, but in actuality it was not the case—especially for DNA molecules that contain components derived from different biological species."[42]

Uncertain of the experiment's success, Cohen assigned it to his research assistant Annie Chang, rather than to one of his postdoctoral students whose career might suffer if the project failed. Chang, who lived

Fig. 4. Herb Boyer in his Department of Microbiology office and lab, UCSF, mid-1970s. In the lab photo, note his running clothes drying in the fume hood. (Photographer unknown; photograph courtesy of Mary C. Betlach.)

in San Francisco, became the conduit between the two labs, ferrying plasmid samples back and forth in her Volkswagen Beetle.[43] In remarkably short order, they had results. On a triumphal day in March, Boyer and Helling examined the electrophoresis gels displaying the various DNA fragments. There in plain sight was a telltale band composed of two types of plasmid DNA standing out in fluorescent orange. To their inestimable joy, they had not only recombined DNA—*they had cloned it!* The engineered plasmids with their ability to reproduce themselves in the bacterial cells had also faithfully cloned the foreign DNA inserted into them.

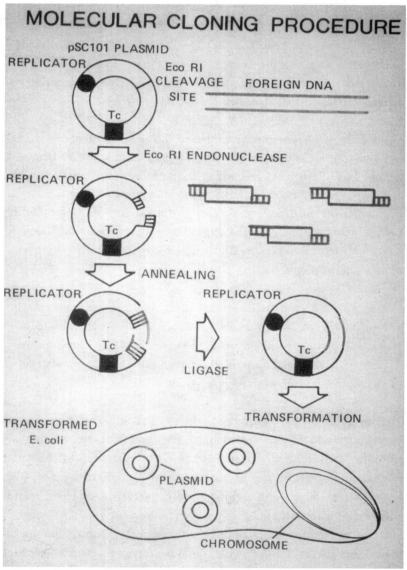

Fig. 5. Diagram of the Cohen-Boyer recombinant DNA procedure. (From a slide courtesy of Stanley N. Cohen.)

The sight brought tears to Boyer's eyes: here before him was evidence of a simple method for isolating and accurately copying specific genes and DNA fragments in virtually endless quantity. He recalled the emotional moment:

> The [DNA] bands were lined up [on the gel] and you could just look at them and you knew . . . [that DNA recombination and cloning] had been successful. . . . I was just ecstatic. . . . I remember going home and showing a photograph [of the gel] to my wife. . . . You know, I looked at that thing until early in the morning. . . . When I saw it . . . I knew that you could do just about anything. . . . I was really moved by it. I had tears welling up in my eyes because it was sort of a cloudy vision of what was to come.[44]

For Cohen, "That moment was elation."[45] The experiment had worked like a charm. He and Boyer had invented a technique that outshone in simplicity and efficiency anything the Stanford biochemists or anyone else had devised for joining DNA fragments. But the crowning accomplishment was the invention of a straightforward technique for cloning DNA, a technique so simple that high school students would soon use it. Boyer later remarked on how neatly new knowledge and breaking techniques had converged:

> Things just came together at that time: the study of small plasmids, transformation of E. coli with [plasmid] DNA, the restriction enzyme business; it was all coming to fruition at the same time. . . . [The experiment] went very fast. It was straightforward. There was not much in the way of struggle. The first experiments more or less worked.[46]

The novel idea for the cloning procedure was strictly their own. But, as both Cohen and Boyer acknowledged, they had benefited enormously from the intellectual milieu and technical breakthroughs in molecular biology and biochemistry for manipulating and characterizing DNA molecules. Scientific and technological convergence set the stage for their dazzling invention.

The paper published in November 1973—one year to the month after the seminal meeting in Hawaii—was a strictly scientific account relayed in the sober language of science.[47] But they allowed themselves one speculation. Although the experiment involved only plasmid DNA,

Cohen and Boyer suggested something more sweeping. Their method, they ventured, was not restricted to lowly plasmids; they foresaw it becoming a general tool for selecting and cloning the DNA of organisms up and down the evolutionary scale.[48] What had started as an experiment to further their respective interests in plasmids and restriction enzymes suddenly appeared to possess far wider applications.

Cohen, a cautious foil to the outgoing Boyer, pressed to keep the research quiet until it was published and their priority established.[49] It was not an unreasonable request, particularly considering the fearsomely competitive Stanford biochemists working a few floors away. But in June 1973, before publication of their experiment, Boyer attended a meeting on nucleic acids, one of the annual Gordon conferences devoted to specific scientific topics. He had struggled for three long months to keep their discovery secret. Innately open and effervescent, Boyer found the request agonizingly difficult to honor. A colleague and friend of both scientists observed, "Herb can't keep anything to himself. Stan, on the other hand, tends to be relatively secretive about what he is doing."[50] Mary Betlach, a technician in the Boyer lab at the time, agreed with the assessment of Boyer: "Herb is the kind of guy that never held anything back.... As soon as something happens, he doesn't care if it's written up; he wants to talk about it."[51]

Arriving at the conference, Boyer found their discovery impossible to keep to himself. Here he was in the company of the very scientists he suspected would thrill to the invention. In a session on restriction enzymes, he explained the amazing properties of EcoRI and the cloning experiment he and Cohen had recently concluded. His presentation provoked a few technical questions, but the method's radical implications fell flat. Only somewhat later did the participants grasp its electrifying significance. The light dawning, one scientist exclaimed: "Well, now we can put together any DNA we want to."[52] He might have added, "And clone it in bacteria." With Boyer's talk, the word was out, carrying far-reaching repercussions for science and beyond.

Cohen, meanwhile, was hard at work. Proceeding without the UCSF group, he and Chang launched a second experiment. The first experiment had dealt with two closely related plasmid species that both inhabit *E. coli* bacteria. In the second, Cohen and Chang sought to determine whether they could join and replicate plasmid DNA from two unrelated bacterial species. Once again, the recombinant plasmids with antibiotic resistance

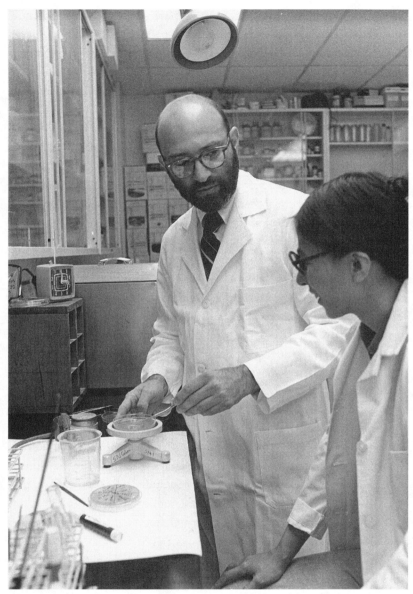

Fig. 6. Stan Cohen and Annie Chang in their Department of Medicine lab, Stanford University, 1975. (Photographer unknown; photograph courtesy of Stanley N. Cohen.)

characteristics of both plasmid parents reproduced in the bacterial cells. Here was evidence that foreign DNA could propagate in bacteria. Cohen proposed for the first time in print what he and Boyer had privately considered: scientists using their technique might implant bacteria with foreign genes for various new and useful functions. He mentioned photosynthesis and antibiotic production as examples of processes potentially possible to introduce.[53]

For the last experiment, the Boyer and Cohen teams reassembled, with the addition of Paul Berg's graduate student John Morrow and Boyer's research partner Howard Goodman. Their aim was ambitious: to determine if the method could clone the DNA of a complex organism. To do the experiment, the team needed a sample of animal DNA. Chance was again on their side. After presenting the cloning procedure at the June 1973 Gordon conference, Boyer had mentioned to Morrow what he considered the obvious next step: to attempt to clone the DNA of a higher organism. The problem was where to obtain a sample of well-characterized animal DNA, a rare commodity at the time. They required DNA of known structure to be able to distinguish it from the welter of extraneous bacterial and plasmid DNA within the bacteria. Morrow told Boyer that he had in a freezer at Stanford some purified DNA from a frog (*Xenopus laevis*) that a mentor had characterized. If his mentor approved, Morrow would provide a sample for the experiment Boyer and Cohen had in mind.[54] It was a stroke of exceptional good fortune. The Cohen-Boyer team now had within their grasp the scarce material it needed. Boyer placed a hurried call to Cohen, telling him the good news. They set up what would be their last collaborative experiment.[55]

Losing no time, they began the experiment in July 1973, using the same overall approach of the previous experiments. Once again, success was resounding: the recombinant plasmids containing the frog DNA faithfully replicated in the bacteria, even though the frog DNA came from an animal many levels higher on the evolutionary tree. The results proved Cohen and Boyer's suspicions to be dramatically correct: their method could reliably clone complex animal genes in primitive bacterial cells. The finding sparked Boyer's sense of humor. A colleague telephoned to learn how they had identified the bacteria containing the frog DNA. "Herb," the colleague reported, "just said he kissed every (bacterial) colony on the [culture] plate until one turned into a prince. Then he hung up, and I had to return the call to get the real answer."[56]

Although Boyer made light of their success, both he and Cohen instantly recognized the experiment's stunning significance. Their procedure gave every indication of working with the DNA of all and every creature regardless of its position in the hierarchy of nature. Up to that moment, the lack of a straightforward method for investigating the genetics of animals and humans had frustrated researchers and held back the field. In the frog work, Cohen and Boyer achieved a long-sought goal in molecular biology: the invention of a simple and efficient method for selecting specific genes from any imaginable organism and accurately reproducing the genetic material in pure and unlimited quantity. In one of the most influential sets of experiments in biology, Cohen and Boyer had flung open a door long shut to the productive study of the genetics of higher organisms. As Boyer remarked: "This technology could take you to the point where you could isolate any gene and make large quantities of it and then study the hell out of it if you wanted to."[57] On another occasion, he put it more soberly: "I think the great thing about this technology was that it was very straightforward, very simple, and it didn't take much to transfer the technology into the laboratory. It just became so widespread in a very short period of time, and an amount of good things came out of it. What more could you ask for?"[58]

In fact, scientists—Cohen and Boyer included—*did* ask for more. Their experiments had not answered a significant question: could simple bacteria "read" the complex genes of higher organisms and express them as proteins like insulin, growth hormone, and so on? No one anywhere had provided a valid and definitive answer.

PATENTING AND POLITICS

That question hung in the balance for later exploration as Stanford officials woke up to the significance of Cohen and Boyer's invention. Their frog DNA research, published in May 1974, prompted Stanford to issue a news bulletin, timed for release the day the publication appeared.[59] The bulletin mentioned the method's utility in basic research and went on to tout its potential in making pharmaceuticals. It quoted Stanford's Nobel laureate Joshua Lederberg opining grandly that the cloning method "may completely change the pharmaceutical industry's approach to making biological elements such as insulin and antibiotics."[60] He was not the first to prophesy the genetic engineering of pharmaceuticals—

such ideas had circulated in molecular biology for some years.[61] But coming from a Nobel laureate and respected figure, Lederberg's opinions counted, helping to plant a concept of the Cohen and Boyer procedure as an invention with sweeping industrial potential. The next day the *San Francisco Chronicle* picked up the story, focusing on the possibility of bacteria being transformed into "factories" for the production of insulin and other drugs.[62] The "microbe as factory" idea also was not new, as humankind's age-old use of microorganisms in making cheese, bread, beer, and wine attests. But the Cohen-Boyer work gave the metaphor new currency that future popular accounts would monotonously repeat. The research caught the attention of *Newsweek*, which carried an article on "the gene transplanters" and their goal of using bacteria to produce pharmaceutical and agricultural products.[63] The message of commercial promise was hard to miss.

A front-page article in the *New York Times* provoked the first concrete step toward commercializing recombinant DNA technology.[64] Alerted to a breaking story, a *Times* science correspondent telephoned Cohen, who recounted the industrial applications he foresaw. The resulting article highlighted the novel technology's likely practical uses in medicine and agriculture.[65] A clipping of the story landed on the desk of Niels Reimers, the enterprising director of Stanford's Office of Technology Licensing. Reimers administered a patenting and licensing program that actively solicited faculty inventions for patenting in a manner new to academia. He read the *Times* article and immediately called Cohen to discuss a possible patent application. The suggestion caught Cohen by surprise. Despite his recognition of the invention's potential practicality, his reaction was to question whether one could or should patent basic research findings.[66]

At the time, biomedical scientists in American universities were seldom preoccupied with patenting and intellectual property protection, even at a university as entrepreneurial as Stanford.[67] Most supported, at least in the ideal, an academic culture valuing open intellectual exchange and the sharing of research results and materials. Although universities throughout the twentieth century had sought patent rights on inventions in practical fields such as chemistry, engineering, and agriculture, patenting in academic biomedicine was controversial on ethical grounds and considerably less frequent—hence Cohen's surprise at Reimers's suggestion.[68] A common belief dating to the early years of the century was that discoveries in biomedicine, especially those related to human health, should be publicly available and not restricted by patents. In some fast talking about patents as a prime means to encourage an invention's commercial development, Reimers managed to overcome Cohen's reservations and persuade him to agree to Stanford filing a patent application on the basic recombinant DNA procedure.

Cohen then approached Boyer, who after a few queries agreed that Reimers should proceed with a patent application. The effort seemed to Boyer a long shot but worth a try. After all, Reimers with Stanford behind him was willing to take the brunt of managing what would become a six-year, politically fraught effort to prosecute the patent. The University of California, as Boyer's employer, then became a cosponsor of the patent application. Cohen spent long hours with an outside patent attorney, struggling to get the wording and claims right. Boyer, in contrast, was quite willing to let others manage the patenting process: "I must admit, I didn't have a lot of patience with patent law and trying to figure it out. So I just told [the patent attorney] everything I knew and the guy went ahead and did it. Stanley helped him out quite a bit; they were always working on it together. I tossed in a few ideas."[69]

The two universities routinely required external evaluations before sinking money into a patent application. One molecular biologist reviewing the Cohen-Boyer application wrote with remarkable foresight:

> This technological development very clearly has immediate applications and probably represents one of the most outstanding new developments in molecular biology in recent years. It is a far-reaching development and has extremely high potential with respect to its commercial application. If the patent is successful there is little doubt that it represents a potential source of considerable amount of royalties for the Universities involved.[70]

On November 4, 1974, Reimers filed a patent application on the recombinant DNA procedure on behalf of Stanford and the University of California. He had beaten the patent bar by a slim week. U.S. patent law requires the filing of a patent application within the year following the invention's first public disclosure, in this case the November 1973 publication on the cloning procedure.

A heated political debate over the possible hazards of genetic engineering gained momentum as the patenting process got under way. In his June 1973 Gordon conference presentation, Boyer had unwittingly planted the seeds for what came to be known as the recombinant DNA controversy.[71] His remark that the experiment with Cohen had created a novel combination of antibiotic resistance genes triggered concern about the safety of the new procedure. A majority of the attendees voted to send letters to the National Academy of Sciences and the National Institute of Medicine requesting formation of a committee to investigate the potential risk of recombinant DNA experiments and the need for research guidelines. The National Academy then formed a committee, with Paul Berg as chairman, to study safety measures for research in the new field.[72] While the committee deliberated, Cohen faced a growing problem—how to handle requests for his plasmid, at the time the only one suitable for use in cloning research. He decided to restrict distribution to scientists planning experiments that he judged would not create new and possibly dangerous combinations of antibiotic resistance genes.[73]

In July 1974 the so-called Berg committee published a letter in *Science* signed by ten prominent scientists, including Boyer and Cohen. It called for a temporary moratorium on certain kinds of recombinant DNA research until a conference convened to consider the risks and develop research guidelines. The *Federal Register* announced in November the formation of the Recombinant DNA Advisory Committee, with a mandate to advise the NIH director on technical matters related to recombinant DNA research.[74] In February 1975 a select group of about a hundred molecular biologists arrived at the now-celebrated Conference on Recombinant DNA Molecules at California's Asilomar conference grounds to consider the technical issue of laboratory research safety, explicitly avoiding deliberation on the technology's larger social and ethical implications. Striving to avoid government regulation, the scientists proposed to devise their own safety regulations with the idea that recombinant DNA research could then proceed. After contentious debate, the participants came up with a preliminary draft of recombinant research guidelines. Cohen, Joshua Lederberg, and James Watson stood out as the only dissenters in the rushed floor vote to approve the draft. Cohen refused to endorse what he saw as a politically motivated document that he and much of the assembly had not reviewed. The conference in his opinion had turned into "a scientific witch hunt" that gave him heartburn and lingering anxiety.[75] Even the more sanguine Boyer was under duress, disturbed by the in-fighting and politicking. He later labeled the Asilomar conference "a nightmare" and admitted he was too upset to sleep.[76]

The Stanford-UC effort to patent the Cohen-Boyer procedure was swept into the swirling political debate over the safety of recombinant DNA research, complicating the patenting process and prompting Cohen's tense vigilance in matters related to DNA politics. In the enflamed political climate, critics—Arthur Kornberg and Paul Berg prominent among them—attacked the two universities and the two co-inventors for attempting through patenting to privatize and profit from a basic biological discovery. Berg was particularly upset, charging that the proposed patent covered "production of all possible recombinants, joined in all possible ways, cloned in all possible organisms, using all possible vectors [vehicles such as plasmids and viruses used to transport genetic material]."[77] Others charged that a successfully issuing patent would encourage the technology's dissemination to industry, carrying along with it a burden of potential hazards. It was an anxious period for both Cohen and Boyer, singled out for criticism as creators and practitioners of recombinant DNA technology and inventors on the patent application. But it was especially troublesome for Cohen. DNA politics, contention over the patent application, and his own cautious nature argued for maintaining a low profile and avoiding further rocking the boat. Throughout this period, he kept a slogan on his desk that said, in effect, if whales don't rise to the surface, they don't get harpooned.[78] It was not a perspective conducive to bold ventures.

STEPS TOWARD COMMERCIALIZATION

The two cloning experiments were the last time Cohen and Boyer ever collaborated. After 1974 they set off on separate tracks. Like most of their colleagues, both subscribed to the utilitarian aims long a theme of American science and were accustomed to justifying their grant proposals in terms of the eventual social utility of the proposed research. The various social movements of the 1960s, in which both Cohen and Boyer had actively participated—Cohen singing and playing political protest ballads on his banjo, and both scientists attending anti-Vietnam rallies—had served to reinforce the notion that publicly funded research, particularly in the life sciences, should lead to practical applications of use to the public.[79] Yet despite this utilitarian strand in American science, biomedical culture into the late 1970s was notably inhospitable to professors forming consuming relationships with business, let alone taking the almost unheard-of step of founding a company without giving up a professorship. Academic cultural tradition, the precarious political context of recombinant DNA research, and the fact that Cohen and Boyer had no desire to leave academia argued against either scientist giving serious consideration to forming a company. No evidence at the time of their seminal research collaboration suggests that they did.

Then, early in 1975, Cohen received an offer that he saw as a means to advance commercialization without upsetting the political apple cart and detracting from his academic responsibilities. Ronald Cape, the smooth-tongued cofounder and chairman of Cetus Corporation—a start-up formed in Berkeley in 1971 to capitalize on an automated method for selecting antibiotic-producing bacteria—offered Cohen a position on the scientific advisory board.[80] "All of a sudden," Cetus cofounder Pete Farley breathlessly put it, "Stan Cohen and Herb Boyer got together and started snipping up DNA and stitching it back together again; we thought, 'Oh my God, just look at that!' So we immediately went after Cohen and signed him up exclusively."[81] Cape's offer was in line with Stanford policy on faculty consulting, designed to bring the university in close touch with industry and stimulate technology transfer.[82] Corporate consulting at Stanford University School of Medicine was fairly commonplace in the 1970s. In fact, Cape had already convinced Joshua Lederberg and other notables to join the Cetus science advisory board. The laureate, who sought out occasions to extol recombinant DNA as a "technology of untold importance for diagnostic and therapeutic medicine," then promoted his colleague and friend Stan Cohen as an obvious choice for advising Cetus on applying the new technology.[83]

A consultant position suited Cohen in terms of career objectives and personal makeup. He was first and foremost an academic—in basic-science orientation and professional aspiration—and anything but the stereotypical risk-taking, freewheeling entrepreneur. According to Stanford policy, professors were to devote no more than thirteen days per academic quarter to consulting.[84] He would continue to give prime attention to his academic responsibilities; yet as a Cetus consultant, he might have a significant hand in moving the recombinant DNA invention toward industrial application. In May 1975 Cohen signed a consultant agreement that paid $7,500 per year, higher than the $6,000 Cetus initially offered.[85] The increase suggests the value the company placed on his scientific and technical expertise. Accepting the position, as well as the Cetus shares that came with it, brought Cohen into formal association with and equity in a company.

As Cohen took a measured step into the corporate world, Boyer remained grounded at UCSF in his consuming academic preoccupations. His burgeoning lab was cloning a "zoo" of various animal genes and struggling to satisfy a deluge of outside requests for the plasmids and restriction enzymes essential for recombinant DNA experiments.[86] Boyer also published profusely (twenty-three publications in 1973–75) and still found time to play soccer with his two young sons. Preoccupied though he was, he never lost the vision of recombinant DNA's promise in pharmaceutical manufacture. In casual moments with colleagues, he relished tossing around ideas about its industrial prospects. Such freewheeling speculation was entertaining, but he had no thought of founding a company.[87] He was a professor and basic scientist—that was his life and aspiration. Furthermore, the university paid scant attention to capitalizing on employees' inventions. Unlike Stanford, UCSF did not have a history of close interaction with local businesses, and the university's patenting and licensing operation was not active or efficient. An institutional framework for commercializing scientific discoveries and examples of UCSF professors launching start-up companies was almost nonexistent.

Yet it was within the realm of accepted academic practice to test in laboratory-level experiments recombinant DNA's possibilities in practical application. A September 1974 memo outlines Boyer's thoughts along these lines.[88] At a meeting with a Stanford licensing office representative to estimate the industrial potential of recombinant DNA, Boyer stated that he foresaw the "immediate" industrial use of the technology in the synthesis of hormones and enzymes. He mentioned insulin as a likely target. Companies, he predicted, would be interested if, through experiments, the technology could demonstrate increased product yields. Boyer thought the technology was ready to commercialize and on his own he tried to interest "at least one drug company" in exploring that possibility.[89] But the unidentified company, he admitted, was not interested. "I wasn't thinking about starting a company. I was just trying to think about how we could get these [pharmaceutical industry] guys interested to take this [technology] and do things."[90]

Before a year had passed, Boyer had taken matters into his own hands and arranged a research collaboration to test the technology's use in pharmaceutical production. In May 1975 he told an interviewer: "I think this [possibility of producing proteins in bacteria] has a lot of implications for utilizing the technology in a commercial sense, that is, could one get bacteria to make hormones, etc., etc. So that's one project that we're involved with."[91] At Boyer's behest, a chemist in Germany had agreed to chemically synthesize DNA fragments for a tiny gene coding for a human hormone (angiotensin II) and deliver them to Boyer by September 1975. Once the synthetic DNA arrived, the plan was for the Boyer lab to attempt to clone the synthetic gene.[92] As the September deadline passed, it became frustratingly clear that the synthetic DNA was not going to arrive. Boyer later learned that the German chemist had decided to attempt the research in his own lab. With no capacity in DNA synthesis at UCSF, Boyer was left without an avenue for carrying his strategy further. As his plan evaporated in the fall of 1975, Cohen, ensconced as a Cetus consultant, appeared to be in a superior position to oversee the first industrial application of recombinant DNA technology.

2

Creating Genentech

> I had these little seeds of thought, fantasy more than anything. But I
> had no idea how you would start a company.
>   Herbert Boyer, April 7, 1994[1]

BOB SWANSON

The impasse that Boyer had reached in exploring the practical application of recombinant DNA technology took an unexpectedly consequential turn in an encounter with an unemployed venture capitalist. Robert
Arthur Swanson was born in Brooklyn, New York, in 1947, the only child
of parents with a year or two of college education. The family moved
when the boy was three to Miami Springs, Florida, a small town near the
Miami airport where his father headed an electrical maintenance crew
for an airline. Bob's childhood was a secure world of Little League baseball coached by his father and the doting attention of his parents and
grandmother. They imbued him from an early age with the notion that
each generation was to do better than the last, perhaps the origin of his
remarkable drive and ambition in later life. He recalled the launching of
Sputnik in 1957 as a landmark in his young life, channeling his interest
toward the science and technology that captivated American boys of his
generation. He attended a large public high school in nearby Hialeah,
where he earned high marks in science and math.

To his family's great pride, Bob entered the Massachusetts Institute
of Technology in 1965, accompanied by a trunk stenciled with his full
name. The trunk would later serve as a makeshift coffee table in his

less-than-opulent bachelor apartments. Bob joined a fraternity and fell right in with its masculine camaraderie. But despite a solid high school education, he found MIT academics stiffly competitive and the course-work a challenge. He received his first D, "a letter of the alphabet," he recalled, "I had never seen before on a grade slip."[2] His fraternity brothers provided the coaching he needed to bring up his grades to A's. A student job as a campus tour leader provided pocket money and a chance to display his outgoing personality. He became chairman of the fraternity's social activities and took to heart its emphasis on brotherhood and interpersonal networking.[3]

Swanson majored in chemistry with the thought of preparing for a career in industry. But a summer job in research at a chemical company changed his mind. "It was a great learning experience," he recalled, "and I discovered a lot about myself. One of the things I discovered was that I enjoyed people more than things. So I said [to myself], 'Gee, this probably isn't going to be what I'd want to do all my life.'"[4] For an individual not given to deep introspection—he was a doer, not a thinker—his self-analysis may have probed no deeper than realizing he wished to be in the thick of things. Swanson then took the initiative to petition MIT to permit premature entry into a graduate business program at its Sloan School of Management while he finished his last undergraduate year in chemistry. He had no time to waste in getting ahead.

Swanson's enterprising move marked a turning point in career direction and, not incidentally, an escape from the draft for the Vietnam War. He found a course on entrepreneurship by far the most eye-opening in the entire business school curriculum. Prophetically, organizational development caught his fancy. The course gave students real-world experience through meeting risk investors, following local start-up companies, and writing practice business plans. He learned that the few professionals in the cottage industry just beginning to be known as venture capital offered private investment funds and management advice to promising entrepreneurial start-ups in exchange for equity stakes in the young companies. Swanson was enthralled: "Here [in venture capital]," he recalled, "you've got ideas going to products at the same time you're building a company."[5] In 1970 Swanson graduated from MIT with an undergraduate degree in chemistry and a master of science degree in management.

Following up on his fascination with venture capital, Swanson took a job with Citibank, which was building a venture investment group in

New York City. In an unusual gesture, the bank gave its newly minted business school graduates a sum of money to invest. The neophytes did very well. Swanson later jokingly referred to the experience as his post-doctoral training in financing and building companies.[6] Impressed, his superiors chose him and a colleague to move west to open a San Francisco office for Citicorp Venture Capital. The plan was to exploit the rich opportunities for risk investment in the Bay Area. Arriving in 1970, Swanson encountered a thriving center of the microelectronics and computer industries in a region thirty miles south of San Francisco, soon to become known as Silicon Valley. It was without doubt the most entrepreneurial region in the world, boasting a refreshingly boundless, risk-tolerant, success-breeds-success culture in which an aspiring young person could spread his wings and try new things.[7] Swanson had found his milieu.

Inevitably, not every Citicorp investment went well. It became Swanson's task to try to salvage the bank's stake in a company rapidly going downhill. Serving on the board of the failing company, he met Eugene Kleiner, who with Thomas Perkins in 1972 had founded Kleiner & Perkins, a venture capital partnership with offices in San Francisco.[8] Taking a measure of Swanson, Kleiner was impressed, according to Perkins, with the young man's ability "to think straight and get things done."[9] When Swanson decided to leave Citibank and seek a new position, Kleiner recommended him to Perkins to fill a vacancy at the partnership. Perkins, a former Hewlett-Packard engineer with a Harvard MBA, respected Kleiner's ability to assess individual character and motivation. Late in 1974 Swanson joined Kleiner & Perkins as a junior partner in its Menlo Park office on Sand Hill Road. He was twenty-six.

One of Swanson's assignments was to monitor the partnership's substantial investment in Cetus Corporation, the company about to acquire Stan Cohen as scientific adviser.[10] Kleiner and Perkins worried that Cetus was not focused on product development and feared their equity stake was turning sour. As part of the effort to get Cetus on track, in 1975 Swanson arranged a luncheon with company cofounders Ron Cape, Pete Farley, and Nobel laureate Donald Glaser. At the last minute, Perkins also came along. The two venture capitalists hoped to persuade management, as Perkins put it, "to do something that would amount to something" and redirect the company along focused and productive lines.[11] The brainstorming session ranged over a number of emerging technologies that Cetus might develop, including the new phenomenon of recombinant DNA.[12] Glaser, a friend of Cohen's, presented a glowing picture of the cloning procedure and its possibilities in genetic engineering. Cape and Farley showed no interest, and Perkins, with no background in biology, failed to grasp the technology's industrial potential. Swanson alone was captivated by Glaser's account. Perkins recalled: "But Swanson *really* got that. I remember the next day he took me aside, and he said, 'This idea [of genetic engineering] is absolutely fantastic; it is revolutionary; it will change the world; it's the most important thing I have ever heard.'"[13]

Swanson spent the next few weeks reading up on recombinant DNA technology and urging Cape and Farley to take it up at Cetus—to no avail. He had more luck with Kleiner and Perkins, his infectious fervor finally convincing them that a technology potentially capable of making medical substances in bacteria had striking industrial possibilities. As Perkins recalled: "We became very interested in gene splicing—all of us, Kleiner, Swanson, and myself. We tried to encourage Cetus to do it. Bob tried very hard. I proposed to Cetus that we set up a separate division of Cetus to do that and put Bob in charge of it. They wouldn't hear of it. It was absolutely rejected. So we reached a dead end with Cetus."[14] In frustration, Kleiner and Perkins sold the partnership's shares in Cetus and abandoned the company.

Cetus was not alone in its hesitation regarding the industrial application of recombinant DNA technology. Pharmaceutical and chemical corporations, conservative institutions at heart, also had reservations, anxious not to lose out if the radical approach proved competitive but also aware of the many unanswered questions concerning its industrial implementation and productivity. In the mid-1970s industry's common watchword regarding recombinant DNA was "wait and see." Only with evidence of commercial feasibility were established corporations willing to consider putting human and material resources into trying to transform the basic-science technique of recombinant DNA into a productive industrial technology. Cetus, despite its entrepreneurial traits, did not begin to build genetic engineering research facilities until December 1976, more than a year and a half after Stan Cohen had become an adviser on commercializing the technology. Only in 1978 did Cetus's state-of-the-art laboratory facilities, designed to contain any imaginable hazard, finally become operational and the first exploratory recombinant DNA experiments were launched.[15] Despite the striking opportunity to become a first mover in a virgin industrial field, Cetus management felt no urgency to apply the new technology. As Cape put it in 1978, "Whatever practical applications I could see for recombinant DNA . . . were five or ten years away, and, therefore, there was simply no rush to get started, from a scientific point of view."[16]

DNA politics presented an additional roadblock. Cape and other corporate leaders hesitated to enter a politically precarious field in which the safety of recombinant DNA was hotly debated and the imposition of government regulation a seemingly likely outcome. In 1976 Congress had a host of bills before it, all stipulating regulation in one form or another of recombinant DNA research. Additionally, a number of cities, from Berkeley to Cambridge, had passed or were considering restrictive local ordinances. That June nearly thirty corporations and two manufacturing associations—all interested in genetic engineering's industrial prospects but wary of attendant political problems, sent representatives to meet with Donald Fredrickson, the director of the National Institutes of Health. The agenda was to discuss the impact on industry of the NIH guidelines for recombinant DNA research, due for formal public release that July. A trade publication reporting on the meeting wrote that the industrialists expressed "no detectable enthusiasm for the guidelines" and quoted one executive as stating that "what are guides today will be regs tomorrow"—regulations, he warned, that would stifle industrial R&D in the new field.[17] Not surprisingly, industry preferred voluntary regulatory compliance or, better yet, no regulation at all. Pharmaceutical companies, weighing the worrisome political issues on top of recombinant DNA's uncertain industrial feasibility, largely decided not to initiate internal programs for the time being.

Calamity meanwhile had befallen Bob Swanson. After the denouement over Cetus, Kleiner and Perkins began to question Swanson's suitability as an associate. "So we reached a dead end with Cetus," Perkins commented. "Kleiner had kind of reached a dead end with Swanson. So we advised Bob that he should seek employment elsewhere."[18] At the end of 1975, Swanson found himself suddenly looking for work, his career plans dashed, no alternative direction in sight, and no one anxious to take him on. Reduced to living on monthly unemployment checks—"$410 a month tax-free"—he scrimped to meet expenses:

My half of an apartment in Pacific Heights [an exclusive San Francisco neighborhood] was $250 [in rent]. My lease payment on the Datsun 240Z was $110, and the rest was peanut butter sandwiches and an occasional movie, plus I had a little savings, not very much at that time.[19]

It was a rude awakening for an exceedingly ambitious young man with high-flying career expectations. Anxious though he was, he kept his options open, considering a dizzying variety of employment opportunities. He recalled:

> I was interviewing a lot of people, from large companies like Intel, where I could get some operating experience before I went out and started on my own, to lots of smaller companies. I probably had three interviews a day for three or four months. This was a pretty scary period.[20]

But all the while he kept coming back to recombinant DNA and his vision of its commercial promise.

FOUNDING GENENTECH

Sometime late in 1975, Swanson decided to act, driven by precarious circumstance and naive faith in the technology's commercial prospects. Recombinant DNA felt to him "like important stuff," important enough to build a company upon.[21] His seven years in venture capital had provided valuable training in raising money and advising new companies, but the experience had also made him feel "like a coach on the sidelines."[22] He wanted a piece of the action; he wanted a company of his own. Culling names from publicity on the 1975 Asilomar conference on recombinant DNA, he drew up a list of scientists prominent in the field. Swanson began to cold-call the scientists, asking if they thought the technology was ready to commercialize. Without exception, all believed recombinant DNA had industrial promise but surmised it would require a decade or two of development before a commercial payoff.[23] Persisting despite the rebuffs, Swanson called Boyer, oblivious of the fact that he was contacting an inventor of the technology. He wanted to start a company, Swanson told the preoccupied professor, and he thought recombinant DNA was ready to commercialize. He was probably right, Boyer replied laconically, thinking to end the call and get back to work.[24] Swanson, not one to give up easily, pressured him for a meeting. Boyer reluctantly gave in. He would see Swanson—but only briefly. "He told me he was very busy,"

Swanson recalled. "He was friendly, but *busy*. He agreed to give me ten minutes of his time on a Friday afternoon."[25]

In January 1976 Swanson, smartly dressed in a suit and tie, thinning hair trimmed and tidy above a youthful face—he was all of twenty-eight—strode purposefully through UCSF's dingy corridors to Boyer's cramped office in the Department of Microbiology. If Swanson had doubts, he didn't show them. His business attire signaled to puzzled lab folks a visitor from the alien world of commerce. Something was up—but what? Short and stocky (a magazine article later described Swanson as not a big man unless standing on his wallet), he was a study in contrasts to Herb Boyer, who was substantial and disheveled, his signature leather vest in keeping with casual academic dress. The professor greeted Swanson in his usual genial fashion but with a thought of dismissing him in short order and getting back to the affairs of the day. As the conversation continued, Swanson learned that Boyer was not only a practitioner of recombinant DNA technology; he was one of its inventors. Still more surprising, he found that Boyer had gone so far as to arrange a research collaboration with a German chemist to test its industrial possibilities. It had not worked out, but he and Boyer seemed of one mind regarding the technology's commercial possibilities. In spite of himself, the reluctant professor grew more and more intrigued. Here before him was an eager young man asserting that he knew how to go about creating and financing a company—unknown and forbidding territory for an academic like himself.

As their conversation continued, the two found themselves immediately compatible, doubtless discovering a bond in their enthusiasm for the technology and in a shared down-to-earth style. They adjourned to a neighborhood tavern, the allotted ten minutes extending to three hours "and at least as many beers."[26] Swanson related his experience in funding and advising high-technology companies, describing to a rapt Boyer the function of venture capital in financing start-ups. Boyer reminisced:

> When Bob came along, he explained venture capital to me. He had this desire to start a company of his own, and he didn't want to start out in the usual fields in the Bay Area at the time, computers or running shoes or other things that were popular at that time. He wanted to do something different. . . . He had read a lot about the [recombinant DNA] technology, and thought it might be useful. I said, "Sure, why not."[27]

As Boyer remarked years later, "You take two naïve people and put them in a room, they just boost each other over the bar."[28] Naïveté, in fact, was a salient quality of their concept for a company. "I always maintain," Boyer reminisced, "that the best attribute we had was our naïveté. . . . I think if we had known about all the problems we were going to encounter, we would have thought twice about starting. . . . Naïveté was the extra added ingredient in biotechnology."[29]

Boyer had cause for quick and casual assent. Here at last was the chance he had sought to test the technology's utility as an industrial process. Assured that Swanson knew how to go about creating a company and would shoulder the brunt of getting it off the ground, why shouldn't he give it a try? He had no intention of leaving the university; if a company indeed materialized, he would treat it as a sideline. He was first and foremost a professor directing a laboratory at the frontier of recombinant science. That priority made Swanson's proposition appealing for another reason. In an era of declining federal support for basic research, Boyer saw the proposed company as a likely source of research contracts and much-needed funding for his lab. Swanson, he recalled, "said he had access to some money, and I thought it would be a good way to fund some postdocs and some work in my laboratory."[30] By aligning himself with a company, perhaps he could mitigate the perennially pressing problem of raising funds for his lab. There was also a personal aspect to Boyer's interest. One of his young sons was on the lower end of the growth curve, yet when tested was found to have normal levels of growth hormone. The boy, the pediatrician assured the concerned parents, would likely reach an acceptable height without treatment, a prediction that turned out to be true. The experience nonetheless prompted Boyer to remark to his wife, Gracie: "You know, we could make human growth hormone; all we have to do is isolate the gene. And this," he remarked in an interview, "was before Bob had gotten in touch with me."[31]

Boyer was primed for an industrial venture of some sort before Swanson arrived on the scene. But he needed someone with business and financial experience to put a vague concept into play. "I had these little seeds of thought, fantasy more than anything," he recalled. "But I had no idea how you would start a company and where you would go, what you would do."[32] For his part, Swanson had found a preeminent figure in recombinant DNA research who was willing to serve as the company's research adviser, recruit scientists, and provide the proposed enterprise with an

Fig. 7. Sculpture depicting the initial meeting of Swanson and Boyer in a San Francisco tavern, with the living originals. The life-size bronze statue by sculptor Larry Anderson sits outside a research building on the Genentech campus. (Photographer unknown; photograph courtesy of Corporate Communications, Genentech, Inc.)

aura of scientific legitimacy. That evening, in the modest confines of a tavern, they agreed to form a partnership, with each contributing $500 toward legal fees. "I can't be sure in retrospect," Swanson remarked, "whether it was my persuasiveness, [Herb's] enthusiasm, or the effect of the beers, but we agreed that night to establish a legal partnership to investigate the commercial feasibility of recombinant DNA technology."[33] The meeting ended with Boyer recommending an atlas on protein structure for Swanson to research and identify small protein hormones for possible synthesis.

Swanson's subsequent market research led repeatedly to insulin, the hormone whose name was currently in the air as a prime target for genetic engineering.[34] Used in diabetes treatment since the 1920s, the hormone extracted from pigs and cows was an essential staple of medical practice, yet on occasion caused allergic reactions in human recipients. The thinking was that human insulin, as a natural product of the human body, would not present such problems. Furthermore, there was strong scientific rationale for choosing insulin. Unlike many proteins of the era, its molecular structure—the sequence of amino acids making up the protein—was known, information critical for devising an experimental strategy for making the hormone. Insulin, moreover, was one of the smaller proteins, only fifty-one amino acids long, suggesting easier laboratory synthesis. Swanson, bringing his business training to bear, found insulin economics impressive. The hormone was an immense and reliable moneymaker for a number of American and European pharmaceutical houses, with world sales greater than $100 million and growing.[35] A ready-made market with a patient population in place counted as notable advantages for a cash-strapped start-up, precluding the daunting expense of promoting and developing a market for a new and unfamiliar product. Yet there was room, Swanson believed, for introducing a human form of insulin that promised to out-compete the animal insulins on the current market. Human insulin, Boyer and Swanson readily agreed, was their hands-down choice for the first industrial trial of recombinant DNA technology.

A target chosen, the overriding need was for financial backing. That was Swanson's province. In March 1976 he completed a preliminary business plan, a critical document for presenting a company to the financial community. After a failed pitch to an heir of a California banking family, Swanson approached his former partners at Kleiner & Perkins.[36] "He was hot as blazes on genetic engineering," Perkins remembered. "He was going to make a career out of this either with us or without us."[37] Swanson presented them with his business plan, a straightforward six-page document. The company's mission, the document proclaimed, was "to engage in the development of unique microorganisms that are capable of producing products that will significantly better mankind. To manufacture and market those products."[38] The goals were notably ambitious, the latter in particular. To become a self-contained company making and selling its own products would entail breaking into the highly competitive and capital-intensive pharmaceutical industry. With the one exception of Syntex Corporation, developer of the oral contraceptive, the industry had not admitted a new firm since the 1920s. Considering this forbidding record, the likelihood of the proposed start-up becoming an independent pharmaceutical company anytime soon was little more than Swanson's grandiose and distant dream.

His short-term strategy was more realistic. The firm, as his business plan laid out, would identify an existing market in which microorganisms manipulated to produce therapeutic hormones could economically compete with older production methods. The company would then

```
                    GENENTECH


                Outline for Discussion
                  KLEINER & PERKINS
                    April 1, 1976



     I.   Corporate Background

     II.  Technology Update

          A.  Synthesis
          B.  Stitching
          C.  Replication and control
          D.  Development Program

     III. Corporate Goals and Strategy

     IV.  Initial Product

          A.  Criteria for selection
          B.  Market Estimate
          C.  Rough economics of production

     V.   Funding Stages and Capital Requirements
```

Fig. 8. Outline of Swanson's presentation to Eugene Kleiner and Tom Perkins, April 1, 1976. (Chief Financial Officer archives, Genentech, Inc., copy courtesy of Robert A. Swanson.)

license the engineered bacteria to one of the established pharmaceutical houses with the know-how and deep pockets to handle the exorbitantly expensive process of drug development, clinical trials, and government review and approval.[39] By partnering with an established corporation, the start-up would not need to acquire the expertise and shoulder the high cost—staggering for a fledgling operation—of the later stages of pharmaceutical production. Swanson then spotlighted insulin as the company's first target, stressing its immense domestic market. He went on to claim that recombinant DNA technology could tap into this gold mine by building "bugs" (bacteria) producing insulin at a fraction of its current selling price and at high profit margins. Swanson requested $500,000 in start-up funds.

Kleiner and Perkins were decidedly intrigued but insisted on hearing from Boyer, the all-important scientific mind behind the proposed enterprise. Perkins recalled the subsequent meeting:

> [Swanson's business plan] was very conventional in that I [representing Kleiner & Perkins] would put up the money, they would hire the people, and it would be a very straightforward venture. I took the view that the technical risk was enormous. I remember asking, "Would God let you make a new form of life like this?" I was very skeptical. I said that I would agree to meet with Boyer. He came in the same week, and we sat down in our conference room for about three hours. Of course, I have a background in physics, electronics, optics, computers, lasers. Biology was never a strength for me. I really didn't know what kind of questions to ask. So I said, "Let's just go through it step by step. Tell me what you're going to do. What equipment you'll need. How will you know if you've succeeded? How long will it take?" I was very impressed with Boyer. He had thought through the whole thing. He had an answer for all those questions—[we'll] need this equipment, these basic chemicals, and take these measurements, and on and on. I concluded that the experiment might not work, but at least they knew how to do the experiment.[40]

Boyer's clear explanation of complex science and its real-world applications did the trick. Kleiner and Perkins agreed to invest $100,000—venture capital's first stake in recombinant DNA technology. But it was only small fraction of the $8 million total in the first-ever Kleiner & Perkins venture fund.

The strategy of the two savvy venture capitalists, as was true in risk investment generally, was to make carefully calculated investments in young companies for the chance of substantial future financial return.[41] However, Kleiner and Perkins, both with operating experience, differed from some of their venture capital colleagues in taking an assiduously hands-on approach to managing their companies, assuming a seat on the board of directors, and advising on corporate management and development. Their mode of operation was a striking departure from the invest-and-leave-it approach of most East Coast financiers. Yet despite careful oversight and intent to reduce risk, the business of Kleiner & Perkins was by its inherent nature extraordinarily speculative. As Kleiner wrote to an inquiring California securities regulator apparently concerned about the risky nature of their Genentech investment: "Kleiner & Perkins realizes that an investment in Genentech is highly speculative, but we are in the business of making highly speculative investments."[42] Yet the risk represented in the enterprise that Swanson proposed was greater than the partners envisioned taking on. Their other investments were in companies with existing products or with products prototyped and in sight of the marketplace. The company Swanson and Boyer proposed lacked products, near term or otherwise. Far more worrying, no evidence existed that they or anyone else could transform the basic-science technique of recombinant DNA into a feasible industrial process, let alone come up with marketable products. Asked later about the level of risk the partnership took on, Perkins replied: "Very high. I figured better than 50–50 we'd lose it. But it's rare when the odds on a new technology are better than 50 percent." Second thoughts? "Not at all. If it worked, the rewards would be obvious."[43]

On the promise of the Kleiner & Perkins seed money, Swanson and Boyer dissolved their partnership and on April 7, 1976, signed incorporation papers creating a legal entity known as Genentech. Boyer had come up with the name, a contraction of *genetic engineering technology* and a vast improvement over Swanson's improbable suggestion of "HerBob." As Swanson recalled, "In one of the flashes of brilliance for which [Herb] is famous, he immediately came up with Genentech. . . . It seemed like a *terrific* name, and the entire process took maybe ten seconds."[44] In the May closing, Kleiner and Perkins handed over a check for $100,000 and acquired 20,000 shares of Genentech preferred stock.[45] Swanson was designated president and treasurer at a monthly salary of $2,500, his anxious period of unemployment finally over. Boyer became vice president and secretary at $1,000 a month. Both men received 25,000 shares of Genentech stock.[46] Their salaries were fairly standard for a start-up of the period. The future value of their shareholdings would be a different story.

Fig. 9. Bob Swanson around the time of Genentech's foundation. (Photographer unknown; photograph courtesy of Corporate Communications, Genentech, Inc.)

As a condition of the investment, Perkins joined Swanson and Boyer on the board of directors and was elected chairman. Little did Perkins know at the outset how heavily instrumental he would continue to be in the company's constant fund-raising. "What was so different about Genentech," he later observed, "was the astonishing amount of capital required to do all this. I know, on day one, if anyone had whispered into my ear that, 'For the next twenty years, you will be involved in raising literally billions of dollars for this thing,' I might not have done it."[47] For almost two decades, Perkins would serve as Genentech's hands-on director, often spending one afternoon a week at the company, year in and year out, deeply involved and bringing his keen operating sense and vast network of connections to bear on the firm's direction and development. Looking back on his service, he recalled, "I can't remember at what point it dawned on me that Genentech would probably be the most important deal of my life, in many terms—the returns, the social benefits, the excitement, the technical prowess, and the fun. By '79 I was a total Genentech junkie. I was committed to making Genentech into a huge success."[48] The executive board's deliberately small size allowed for quick decisions, unanimity, and flexibility at the top. Firm direction and product focus, rather than star power, were the company's priorities.

Of even starker economy was Genentech's scientific guidance. Boyer was the company's main scientific adviser, and a part-time one at that. In this regard, the company was an anomaly in the research-based business community, especially in comparison with Cetus's star-studded board of scientific advisers. Yet in Boyer existed a fund of rare knowledge and technical expertise, access to scientists proficient in essential technologies, and ties to and credibility in the academic community, not only at UCSF but throughout molecular biology. He was the increasingly well-known chief of a lab at the leading edge of recombinant research, with all the conceptual and technical knowledge and human and material resources that entailed. In December 1975 Boyer had resigned from the "more or less sterile environment" of the microbiology department and joined the faculty of the increasingly powerful Department of Biochemistry, where his interests in genetic manipulation and molecular events were closely aligned.[49] UCSF's administration at last allotted him generous, up-to-date lab space in one of the research towers and an appointment in biochemistry's new Division of Genetics. He was now located in one of the key departments in molecular biology/biochemistry worldwide. The Boyer lab, with room to expand, was a beehive of recombinant DNA research and technical activities—in plasmid construction and preparation of the enzymes required for gene splicing and cloning. It had become a virtual factory of materials needed for recombinant research, sending them out free of charge to researchers worldwide at a time when commercial reagents were hard to come by. For additional technical expertise, Boyer could turn to his research partner Howard Goodman and his lab of young scientists adept in genetic sequencing. Boyer's appointment in a leading department of recombinant DNA science and the network of personal, intellectual, and material resources he commanded were to be of inestimable value to Genentech. In July 1976 Boyer attained the rank of full professor.

LEGAL AND POLITICAL OBSTACLES

The skeleton of a company established, Swanson had a number of pressing concerns to attend to. Among the foremost was to petition Stanford and the University of California for a license to practice recombinant DNA technology under the patent that the universities were pursuing. If a patent was issued, the universities would demand that Genentech and any other company using the technology purchase a license—or risk being sued if they did not. Swanson also appreciated that owning a license on an important patent would add credibility and luster to his fledgling firm, making it more attractive to investors, or so he hoped. Beginning a few weeks after Genentech's incorporation in April, Swanson insisted in a flurry of letters and visits to the patent licensing offices at Stanford and UC that obtaining license rights was critical to the firm's survival. But it was not just any license he sought; he set out to negotiate an exclusive license to make recombinant pharmaceuticals and gain control over that specific application.

In April 1976 Swanson presented Niels Reimers of Stanford's Office of Technology Licensing with a business plan, written to play upon Stanford's and Reimers's interest in prompt industrialization of its licensed technologies.[50] The document introduced Genentech—scarcely more than a concept at the time—as poised to turn recombinant DNA into a productive industrial process. Swanson made the case for insulin as the company's first target and his strategy to sell insulin-producing bacteria to a major pharmaceutical company for final development and manufacture. He then reiterated his vision to build a company manufacturing and marketing its own engineered pharmaceuticals. Swanson intended these assertions, all considerable leaps of faith, to remove any taint of risk from an endeavor that at every level was deeply risk-laden. As a final incentive, Swanson offered Stanford and UC four thousand shares of common stock if they agreed to grant Genentech an exclusive license to practice recombinant DNA under the patent, assuming it issued.[51] Cetus president Pete Farley, somehow getting wind of Swanson's offer, called it a "preposterous proposal" intended, as Swanson doubtless had in mind, to tie up an emerging industrial area and gain a substantial competitive advantage.[52]

But it was not only the venture's business risks that Swanson knew would concern Reimers. The national controversy over the safety of recombinant DNA research had grown steadily more turbulent, with Stanford and UC in the eye of the political storm. Not only were the two campuses active centers of recombinant science, but critics also angrily challenged the universities' effort to make patent claims on the method itself. As an invention of publicly supported basic science, they pointed out, it should by right be freely accessible to whoever wished to use it. Furthermore, a patent, should it issue, was designed to foster commercialization of what in this case was deemed a potentially dangerous technology. Recognizing a possibly explosive situation in the making, the two universities were tiptoeing through a public-relations minefield, anxious not to trigger further outcry against the controversial patent application but at the same time anxious to capture licensing fees and royalties if the government granted a patent.

Keenly aware of Stanford's tense embroilment in DNA politics, Swanson added a section to the business plan that addressed Genentech's status regarding the NIH guidelines for recombinant DNA research, not yet formally released but available in advance copies. The guidelines applied only to research funded by the federal government—not to privately supported research such as that Genentech contemplated. But Swanson recognized the political and persuasive advantages of satisfying Reimers and the two universities that the firm's research would be safe under any circumstance. The company, the business plan asserted, did not intend to employ the allegedly dangerous organisms subject to the pending NIH guidelines: "None of the [NIH-]prohibited experimentation will effect [sic] Genentech's work." "Genentech's engineering," the document continued, "will be done in facilities equivalent to medium to low ranges of [biosafety] containment, well within the standard university facilities. No special containment procedures will be needed."[53] Genentech, Swanson was keen to establish, would not give an academic community and agitated public an added cause for worry.

Swanson's salesmanship failed to persuade Reimers. The embattled licensing officer and his university superiors had not decided upon a final licensing strategy that would be both equitable and politically expedient in light of the recombinant DNA controversy. Reimers was of no mind at that point to grant Genentech or any other company a license on recombinant DNA technology. Yet, as he admitted privately, he appreciated the consequences for Genentech if Stanford decided upon a nonexclusive licensing policy: "This [decision] of course means that Genentech will not obtain its desired exclusive [license], that we [Stanford and UC] forgo equity and a possible substantial front payment for an exclusive, and it may mean that Genentech as a viable company cannot survive."[54] Was Swanson discouraged by the outcome?

> No, [he later asserted], because it was pretty clear that someone else wasn't going to get [an exclusive license] if we weren't, and that Niels would make a decision to make [the licenses] broadly available. Remember, at that time nobody else believed [recombinant DNA technology] could work [commercially]. We hadn't even proven that we could make a useful product out of it. So he saw the potential and I saw the potential, but there weren't a bunch of other companies clamoring to invest money in this field.[55]

Perkins, who had previous ties with Reimers and his staff, was also optimistic that Stanford, a notably entrepreneurial university at a time when most institutions of higher learning were not, would eventually grant Genentech a license.[56]

Swanson remained outwardly unflappable despite the licensing setback and worrisomely unfavorable political environment. He was determined to get Genentech up and going, whatever recourse it meant taking. Early on he and Perkins briefly considered moving Genentech offshore in light of "the uproar over genetic engineering in the press and in the public,"[57] as Perkins put it. For undisclosed reasons—but likely related to the advantages of proximity to Boyer and a technically qualified workforce—the two decided that the company would remain in California. Swanson's unswerving—some called it stubborn—determination and persistence were plainly evident. As a colleague remarked, "Bob, like many entrepreneurs, was very single-minded and very goal-oriented, and having set himself in a certain direction, he trudged through whatever was required to attain it. . . . I think one of Bob's great geniuses was his ability to stay the course toward discrete objectives."[58]

Toward the end of 1976, Swanson geared up for a second round of private financing. It was high time. In December Genentech reported total assets of $88,421 and a net loss of $88,601.[59] Swanson, Boyer, and Perkins prepared the first full business plan, a 46-page document.[60] Boyer wrote the technical sections, and Perkins, as lead investor in the second financing round, vetted and doubtless contributed. The plan presented essentially the same corporate strategy and goals laid out in the earlier documents. But there were some differences and considerably more detail. Describing the industrial uses of recombinant DNA as multiple and far-reaching, Swanson at his promotional best told of its sweeping potential in various worldwide markets and Genentech's capacity to exploit it: "With Genentech's technology, microorganisms could be engineered to produce needed protein to meet the world food needs or to produce antibodies to fight viral infection. Any product produced by a living organism is eventually within the company's reach."[61] Although the company planned to start off modestly as a supplier of engineered bacteria to established corporations, its future destiny, Swanson boldly reiterated, was to enter the pharmaceutical industry as a fully integrated, self-sufficient, and independent company.

Swanson went on to assert with more than a little exaggeration that "each stage of the technology has been committed to practice" and that "technical risks have been all but eliminated."[62] The document portrayed a company on the verge of productivity, nowhere alluding to its actual fledgling state or to the fact that its basic technology had never been tested as an industrial process. It did not mention the hostile national environment for a company premised on recombinant DNA technology nor the failure to license recombinant DNA technology under Stanford and UC's hoped-for patent. Like most business plans, it was a thoroughly promotional creation, designed to convince potential investors of the certain success of the business venture and glossing over the harrowing technical, financial, commercial, and political challenges. A timeline for product development listed the first recombinant therapeutic as scheduled for production by mid-1977, a mere six months distant. The development schedule sent an intended message that Genentech, if not already in production mode, was close to it. A diagram of an insulin molecule, its entire sequence of amino acids spread impressively across one full page, suggested a needed product in the making. The plan entirely ignored the fact that the firm had not initiated research on insulin. The optimistic forecasts were typical of entrepreneurial promotion but strikingly out of line with sober reality. Genentech at this point was no more than a virtual company. It had no laboratories, no research equipment, no scientists of its own, no money for sustained development, no patents or licenses securely in hand, no certainty of the impact of the festering recombinant DNA debate.

But there had been one propitious development. The December business plan announced the formation of a "core development team" that included Boyer and his UCSF lab and scientists at two Southern California institutions. Both groups were under contract by Genentech to perform its first experiment. The company, in the contemplated research, was preparing to cross into virgin scientific, technological, and industrial territory.

3

Proving the Technology

... a scientific triumph of the first order.
Philip Handler, President, National Academy of Sciences,
statement in the U.S. Congress, November 1977[1]

Swanson's experience as a venture capitalist had centered on young Silicon Valley companies, each with products that had been prototyped and were nearing or on the market. Genentech presented a very different situation. Its fundamental technology was raw and industrially untested, and it lacked a single product in the pipeline or even on the near horizon. Moreover, Boyer and Swanson seemed to blithely disregard the fact that no one anywhere had succeeded in expressing a foreign protein in bacteria, let alone a protein of insulin's complexity. That was the leap they expected Genentech to take—to shoot straight for human insulin, the ballyhooed substance with proven therapeutic value and an immense worldwide market. Instead, Swanson found that he had to stand by, more or less helplessly, as Boyer and the contract scientists came to a different approach, a different decision, a different target. To his consternation, scientific rather than business considerations drove the change of direction.

A previous collaboration between the Boyer lab and two Southern California scientists provided the rationale, personnel, and material resources for Genentech's first experiment, the critical proof-of-concept research that would determine whether or not its core technology was workable as an industrial process. Genentech's future, as Swanson knew with chilling anxiety, rested in the hands and minds of these scientists.

Results of an earlier experiment arriving early in 1976 convinced Boyer that Genentech was on the right track. He and his lab had collaborated with Arthur Riggs, a quietly inventive molecular biologist, and his colleague Keiichi Itakura, a Japanese organic chemist of Samurai lineage, both at the City of Hope National Medical Center in Duarte, California, near Los Angeles. The collaboration sprang from a seminar Boyer gave at City of Hope in 1975 on the recombinant DNA procedure. In the course of the presentation, it struck Riggs that the novel procedure might solve a problem he and Itakura had encountered in collaborating with a high-powered Caltech group trying without success to crystallize a protein (the lac repressor) of major interest in molecular biology. Itakura had a rare proficiency in chemical DNA synthesis, the laboratory construction of artificial genes and DNA fragments from chemicals off the shelf. But due to the inefficient and error-prone synthetic chemistry of the era, he could make only minuscule amounts of synthetic DNA and with inherent impurities to boot.

The quantities Itakura managed laboriously to synthesize were inadequate for most experiments, including the one Riggs, Itakura, and colleagues were currently conducting. Boyer recalls that after his seminar he and Riggs retired to Riggs's office to discuss an idea for a collaborative experiment that, if they were fortunate, would produce an unlimited supply of pure DNA. Boyer re-created their thinking: "We said, 'Why don't we clone this chemically synthesized DNA and you only have to synthesize it once and we'll let the bacteria do it after that [via the cloning procedure].'"[2] They decided that Itakura would provide a piece of synthetic DNA (the lac operator) he had previously made in his dissertation research, and the Boyer lab would try to clone it. For Boyer, the idea of using synthetic DNA in experiments was not only familiar but one he had put into actual practice. As early as 1974, his lab was using short pieces of synthetic DNA (so-called linkers) in studies of the interaction of restriction enzymes with DNA.[3] Boyer and Riggs agreed on the spot to collaborate.[4]

Boyer returned to San Francisco and assigned the work to Herbert Heyneker, a postdoctoral student who had joined the lab in the fall of 1975, more than eager to take on a hot project. The wiry, boundlessly enthusiastic Dutchman had floated into the lab on a cloud of goodwill. He had brought along with him from Holland a thermos filled with vials of enzymes purified for his thesis research. These enzymes were not then available commercially and had to be begged or borrowed through a network of material exchange or tediously synthesized and purified on one's own. Heyneker's gift was consequently a boon while it lasted for the array of genetic recombination and cloning experiments ongoing in the Boyer lab.

The collaboration proceeded according to plan: Itakura contributed the synthetic DNA fragment, and Heyneker managed to splice it into plasmids and clone it in bacteria. Neither component was easily achieved. DNA synthesis and DNA cloning in the mid-1970s were both painstakingly exacting and time-consuming procedures, each step requiring improvisation and laborious chemistry. The experimental results, arriving in February 1976, exceeded all expectations. The chemically synthesized DNA was biologically functional. The proof lay in the fact that the synthetic DNA bound to a bacterial protein, leaving a telltale sign of blue coloring, an indication of natural biological function.[5] Man-made genetic material had behaved in identical fashion to natural DNA. It was an astounding discovery, blurring the boundary between the chemically inert and the biologically active. Heyneker recalled the thrill: "The most exciting moment was when we demonstrated that we could immortalize synthetic DNA [through cloning] and it becomes part of a plasmid, becomes part of a biological system."[6] He went on to note the psychological impact on colleagues at the outset of their careers:

> We were young, and when you are successful, it helps enormously with your whole state of mind. It helps with your confidence; it helps with the publications you write; it helps with your future, with your career. So it really was a very positive time from that point of view.[7]

Press coverage of the breakthrough quoted an ecstatic Boyer brimming with optimism. "We've gone out of the area of basic science into the area of practical application," he proclaimed, and went on to describe enthralling images of bacterial "factories" efficiently spewing out quantities of insulin, growth hormone, and other pharmaceuticals.[8] DNA synthesis, he was convinced, was the companion technology that would make recombinant DNA a feasible industrial technology, not sometime off in the future, as most imagined, but in the tantalizingly reachable near term.[9] "I thought," Boyer recalled, "it was very important that we [at Genentech] have chemically synthesized DNA, no matter what we did. I thought that was the key."[10] Yet encouraging as the results were, the experiment had not settled a point of critical importance: where was the evidence that recombinant DNA and DNA synthesis could produce functional proteins in bacteria? A number of molecular biologists were highly dubious. Could artificial genes, inserted into bacteria, program them to become microscopic production sites spewing out proteins of human specifications? Genentech's first research project, a make-or-break experiment, would attempt to answer that question. The company's future hung in the balance.

SWITCHING TARGETS

For Itakura and Riggs, the ambition to make human insulin was familiar territory. Since his student days in Japan, Itakura had envisioned building a gene with practical significance. "That was my dream, making a gene," he reminisced. "So maybe insulin gene—or whatever gene."[11] Over the several years he and Riggs had known each other, they had repeatedly talked of chemically synthesizing a gene with biological function. A man-made gene for insulin was at the top of their list. Then Riggs abruptly changed his mind. He attended a seminar in which an endocrinologist presented his work on a number of brain hormones, including one called somatostatin. It struck Riggs that somatostatin, composed of a single chain of fourteen amino acids in contrast to insulin's double chain of fifty-one, was a far better choice for the time-consuming process of chemical DNA synthesis. He and Itakura decided to give it a try.

Early in 1976 they got together to compose on paper a structure for an artificial gene—a gene coding for somatostatin. But their aim was not to design a gene as it existed in nature; they suspected that bacteria near the bottom of the evolutionary tree would be unable to "read" instructions encoded in the genes of higher animals. Instead, their strategy was to construct an artificial copy of a gene whose customized instructions the bacteria would find compatible and express as somatostatin. Riggs explained:

> We didn't try to copy a human gene and put that in bacteria. We designed
> a gene that would work in bacteria. So this was a totally man-made—not

only man-made but man-designed—gene, and I thought that was one of the most key aspects of what we were doing, of our approach. We didn't know that it was going to be successful, but we thought we had a good chance.[12]

Despite Riggs's optimism, he did indeed worry about genetic incompatibilities among species. Like many molecular biologists, he suspected that differences might well exist in the regulation of gene expression in simple and complex organisms. Could bacteria interpret the genes of higher creatures and express them as proteins? No one had a definitive answer. Many scientists were highly skeptical. A prominent molecular biologist wondered whether bacteria were indeed capable of interpreting animal genes engineered into them. "Most molecular biologists," he concluded, "would guess not."[13] Riggs recalled the uncertainty:

So we had shown [in the collaboration with Boyer's lab] that the DNA would function in vivo. Nobody had shown that you could actually make a protein product [in bacteria]. So DNA makes RNA makes protein—none of that had been done using synthetic DNA, or even using any other approach, really.[14]

The main objective of the somatostatin experiment was to determine whether bacteria could in fact express an artificial gene as a functional mammalian protein. A conclusive result would count as a significant contribution to an understanding of gene regulation and expression. It would be of arresting significance for Genentech, whose future rested on its capacity to produce useful proteins in bacteria.

To conduct the experiment, Riggs and Itakura needed funding. In February 1976 they submitted a grant application to the NIH entitled "Human Peptide Hormone Production in *E. coli*."[15] They asked for $400,000 for a three-year project to make somatostatin using DNA synthesis and recombinant DNA technologies. They went on to state with notable confidence, considering the uncertainties involved:

The work proposed here will lead to the production of human hormone peptides in *E. coli*. We think that *E. coli* can be used to produce human hormones more cheaply and of better quality than can be made by synthetic peptide [protein components] chemistry. The availability of inexpensive, high quality human hormones will have many clinical application[s].

Despite these assertions, Riggs later maintained that commercialization was not their immediate goal: "None of the group here was thinking commercial yet—no. But thinking about some demonstration that bacteria could be used as factories to make proteins—yes."[16]

While waiting for the NIH to respond, Riggs received a call from Boyer, who suggested a second collaboration, this time to attempt to synthesize human insulin. Riggs was immediately intrigued. Boyer's offer was a chance to take the technology of their first collaboration a significant step further, to actual bacterial production of a protein. But he took exception to targeting insulin. Somatostatin, he advised, was the substance they should go after. If they succeeded, they could then attempt the more ambitious experiment of making human insulin. Boyer was easily persuaded; a stepwise approach made sense. He agreed to shift the target to somatostatin. No need to worry about money, Boyer explained; Genentech was prepared to fund the research. After Boyer's assurance that despite corporate sponsorship they would publish the results, Riggs and Itakura signed on. It was a fortunate decision. That fall the NIH turned down their grant application. The reviewers decided that Riggs and Itakura could not accomplish the proposed research in the stipulated three years and labeled it "an academic exercise" without practical merit.[17]

Convincing Swanson to switch targets was considerably more difficult. Digging in his heels, he clung stubbornly to insulin, a proven therapeutic substance of world renown and an established moneymaker. He adamantly resisted going after somatostatin, an obscure hormone without clear clinical application and market potential. He wanted products—*marketable* products. "I fought that [proof-of-principle experiment] like the devil because I always hated the idea of doing a demonstration of anything," he recalled. "If you are going to go for something, go for the real thing."[18] The "real thing" was human insulin, the high-profile product that would attract investors and bring in the money Genentech desperately needed. The scientists, on the other hand, focused on experimental do-ability rather than marketability and the bottom line. It was not only a battle of wills; it was a contest between scientific and business objectives—a conflict that research-driven companies repeatedly experience. Riggs recalled:

> I wound up talking with Robert Swanson, so I do know how difficult it was
> to convince him that we should make this small peptide [somatostatin]

that did not have any proven commercial value. Swanson was very worried about that. I had no real understanding [at the time] of how precarious the funding was, and how short a time that Swanson believed we had in order to perform. He eventually agreed that the strategy was to establish feasibility, submit patents, and try to publish the work. So he finally decided that since we could do somatostatin quicker [than insulin] and with a higher chance of quick success, that was the way to go. So he finally agreed but we had to talk to him quite a bit. I think Herb [Boyer] did most of the arm-twisting.[19]

They would go for somatostatin, Swanson grudgingly agreed, but only as a way station in the quest for human insulin.

NEGOTIATING RESEARCH AGREEMENTS

With the scientific teams lined up, Swanson pressed for arrangements to move swiftly so the research could begin. With the firm's survival and his own career at stake, he later admitted, "I was the one that was in a hurry."[20] High on Swanson's list of priorities was establishing research agreements with the University of California and City of Hope. The agreements would provide a legal framework in which the contract research could proceed and would also outline the parameters for intellectual property protection and royalty payments. Soon after Genentech's incorporation, Swanson, working mostly without an attorney to avoid legal fees, began to negotiate with the university. In August 1976 he and two university representatives signed an agreement that Genentech would provide $35,000 to fund research in Boyer's lab, including the salaries of two postdoctoral researchers and university overhead, for the bacterial synthesis of various proteins, somatostatin and insulin in specific.[21] Following standard UC policy, the university would hold title to and earn royalty income on any resulting patents. Genentech would receive an exclusive license under those patents and pay royalties to the university on products sold.

To negotiate a research agreement with City of Hope for the DNA synthesis portion of the research, Swanson decided to secure the services of a patent attorney in Los Angeles. He eventually found Thomas Kiley, a partner at Lyon & Lyon, one of the largest intellectual property firms in the country. As Kiley the wit and raconteur tells it, Swanson called a

GNE CFO files   Somatostatin/insulin

```
                    SPONSORED RESEARCH AGREEMENT


              THIS AGREEMENT, effective this first day of August,
        1976, by and between Genentech, Inc. and the Regents of the
        University of California
              WITNESSETH:
                          1.  BACKGROUND
              1.01  Genentech, Inc., (hereinafter "GENENTECH") is
        a corporation organized and existing under the laws of the
        State of California, havings its principal place of business
        at 475 Sansome Street, 15th Floor, San Francisco, California.
              1.02  The Regents of the University of California
        (hereinafter "Regents") is a nonprofit corporation organized
        and existing under the laws of the State of California, and
        having its principal place of business at Berkeley, California.
              1.03  Regents has available laboratory facilities and
        personnel at its San Francisco Campus (hereinafter ("UCSF")
        qualified in the use of restriction endonucleases, and other
        enzymes, the manipulation of plasmids and other segments of
        DNA, and the expression of DNA that could eventually result in
        the "in vivo" synthesis of various polypeptides, including
        synthesis of somatostatin and insulin.  GENENTECH desires to
        fund research relating to technology for the manufacture of
        such polypeptides and, pursuant to paragraphs 3.03 and 4.01
```

Fig. 10. First page of the August 1976 agreement between Genentech and the University of California. (Chief Financial Officer archives, Genentech, Inc., copy courtesy of Robert A. Swanson.)

Lyon & Lyon attorney and explained Genentech's business: "My partner said later he held the phone away from his ear while Swanson was speaking on the esoteric subject of gene-splicing and said to himself, 'Let's see. Kiley represents lots of these weirdos; I'll send Swanson to him.'"[22] Kiley's previous representation of Miss Nude American and other oddball cases had endowed him with a quirky reputation among his associates. Swanson followed the lead and contacted Kiley. But he insisted that

Riggs accompany him to meet the young attorney and decide whether he and the scientists were likely to work comfortably together. Swanson considered the compatibility of attorney and scientists essential for full disclosure of the technical information necessary to write strong patent applications. Easily convincing Riggs that they could work productively together, Kiley then negotiated an agreement between Genentech and City of Hope that, in notable contrast to the University of California agreement, gave Genentech exclusive ownership of any and all patents based on the work and paid the medical center a 2 percent royalty on sales of products arising from the research.[23]

Kiley turned out to be flexible and attuned to a young company's needs. To save the financially strapped start-up the cost of a hotel room, he agreed on his trips to San Francisco to sleep on Swanson's sofa, provided his host treated him to a decent dinner. For more than a decade, representation of Genentech would give Kiley a venue for his flamboyant legal intellect and veneration for all things technical. It would fall to him to indoctrinate the firm's future scientists, predictably naive in business matters, on the central importance of patenting, especially critical for a fledgling company in which the basic "products" were their own technical know-how and innovative power. It was crucial, then, to protect and monetize this intellectual capital as patents and licenses. But it was not merely the sweet seduction of Genentech's technology and the novel legal issues that drew him. Kiley was also impressed with Swanson, especially his seemingly unflappable self-confidence despite formidable odds.

> I admired his chutzpah in aiming with this controversial technology to mount barriers to entry that stood high around the pharmaceutical industry; admired his moxie in doing that on what seemed a very small amount of money—so small that I did everything I could to spare him expense. . . . Genentech [for me] was becoming a labor of love, if you will.[24]

Genentech's third institutional collaborator was an afterthought. Wishing to cover several bases regarding sources of synthetic DNA, Swanson enlisted Richard Scheller, a graduate student at the California Institute of Technology whom Itakura had schooled in DNA synthesis. That meant forming a sponsored research agreement with Caltech. The document specified that Genentech would reimburse Caltech the $4,000 already expended on Scheller's work, plus any additional funding needed to complete his synthesis of somatostatin DNA.[25] The twenty-two-year-old would receive a monthly stipend of a few hundred dollars and around fifteen hundred Genentech shares. He appreciated the extra money but not the stock certificate. More than two decades later, Scheller, by then a Genentech vice president and director of research, explained his cavalier attitude: "I had a ponytail halfway down my back. I smoked marijuana every day. I didn't give a damn about money or stock or anything. I was a scientist. I didn't, fortunately, throw [the certificate] away, but it really didn't mean anything to me."[26] Scheller's attitude was a result in large part of youthful arrogance. But it also reflected a pervasive inattention or even ignorance to matters of stock ownership and equity participation among biologists at the time.

Decades later the Genentech–City of Hope contract spawned a tangle of interrelated lawsuits that displayed the legal complexities and conundrums emanating from the firm's earliest transactions. City of Hope claimed in a case initiated in 1999 that it was entitled to royalties above and beyond the $285 million on sales of insulin and growth hormone products it had received from Genentech through 2000.[27] The medical center contended that it was due additional royalties from the twenty or more patent licenses arising from the insulin and growth hormone work that Genentech had concluded with third parties. City of Hope calculated that it was due unpaid royalties in the $500 million range. Genentech argued that City of Hope was entitled to royalties only on insulin and growth hormone, the two proteins the medical center had co-developed with Genentech. (Somatostatin never became a commercial product and hence did not figure in the royalty dispute.) The case ended in a mistrial in 2001, with the jury deadlocked 7–5 in Genentech's favor. The following year, in a stunning reversal in a new trial, a jury awarded City of Hope $300 million in compensatory damages and another $200 million in punitive damages for Genentech's alleged concealment of the additional licensing agreements. Genentech appealed, but a judge upheld the damage amounts.[28] In 2008 the California Supreme Court struck down the award of punitive damages on technical grounds and reduced Genentech's total damages from $500 million to $300 million.[29] It was nonetheless a sad ending to a partnership of two institutions that, as the press put it, "gave birth to the biotech industry."[30]

Early in 1977 Swanson launched a second financing round. Kleiner & Perkins took the lead and invested a second $100,000, an inducement for five other funds to invest. By February Swanson had raised around $850,000.[31] The somatostatin research could begin. That month Itakura and Tadaaki Hirose, a Japanese postdoctoral student at City of Hope, began the chemical synthesis of the eight fragments that were to make up the artificial gene. Scheller, a neophyte in such matters, failed to make sufficiently accurate DNA and dropped out of the project.[32] Assisted by Hirose, Itakura worked twelve-hour days, six days a week, struggling with the cumbersome chemistry of the day to make accurate and reasonably pure synthetic DNA.

Boyer, meanwhile, had problems of his own. As director of the lab that was to conduct the molecular biology component of the somatostatin project, he had to deal with the NIH guidelines, formally issued in July 1976.[33] Although they applied only to recombinant DNA research receiving federal funds—and the somatostatin research funding came from Genentech—the UCSF administration, ultra-cautious in light of the political fray, had decided that its new biosafety committee would review all experiments involving recombinant DNA, regardless of funding source. Although the biosafety committee determined the experiment low risk and performable under normal conditions, it stipulated higher physical and biological containment levels, apparently trying to avoid any chance of public criticism. The decision meant that Herb Heyneker would be required to perform the cloning work in the biochemistry department's new biohazards containment lab, recently constructed for conducting experiments the guidelines designated as possibly dangerous. In February 1977 Boyer signed a document holding himself responsible for his lab's adherence to the guidelines in the somatostatin research.[34]

In May 1977 the DNA chemistry component took a turn for the better. Roberto Crea, an avid young Italian from the southern province of Calabria, arrived at City of Hope to begin a postdoctoral fellowship, which he discovered that Genentech entirely supported.[35] Crea had just completed a fellowship in organic chemistry in Holland, where he had had a central hand in developing a more efficient method of DNA synthesis. Although taken by the sunstruck Californian landscape, he wasted no time in getting down to work, enticed by a project to make an artificial gene but knowing nothing about recombinant DNA technology, not even the term. Struggling to communicate in his faltering "Dutch English" with a decided Italian lilt, he learned from Itakura, likewise not a native English speaker, that synthesis of the somatostatin gene fragments was not going well, particularly in regard to one recalcitrant fragment. Crea immediately set to work to construct a new set of DNA fragments, which he purified using a newly purchased apparatus (a high-performance liquid chromatographer) that he alone in the lab knew how to operate. Within weeks Crea had constructed new gene fragments of improved accuracy and purity.[36]

Swanson all the while was breathing down everyone's necks, nervous, close to neurotic, with so much at stake. He went regularly to UCSF to interrogate Heyneker, who remembered Swanson being "extremely impatient and extremely anxious for us to get results."[37] Swanson also flew down periodically to check on the City of Hope scientists. Riggs and Itakura resented the monitoring. "[Bob] was very intelligent and knew just enough chemistry," Riggs remarked, "to ask questions that were kind of annoying."[38] Itakura found the oversight irritating, particularly when it came to Swanson's projected timetable for the research: "[Swanson] has all kinds of schedule—DNA synthesis finish such and such; [somatostatin] expression such and such; and then fund-raising, such and such. He showed me that kind of a table, exactly scheduled, step by step. . . . I said, 'Meaningless.' I told him you never know when experiment works [or not]."[39] Biological research, Itakura meant, did not advance in the predictable and timely manner a business mind expected.

The next step was for Heyneker to stitch the synthetic DNA fragments together to form an artificial gene, using the lab's rare kit of enzymes.[40] Then the Boyer lab's plasmid expert, Francisco Bolivar, constructed plasmids designed to express somatostatin. The young Mexican scientist attached the synthetic somatostatin gene to a bacterial "promoter" gene regulating protein expression and a genetic sequence coding for a small portion of a bacterial enzyme. He then inserted the plasmids into bacteria. The idea was that with the promoter gene turned on, the bacteria would express the somatostatin protein. It would count as a stunning breakthrough if protein expression worked. Anticipating success, Riggs invited Boyer, Swanson, Heyneker, and Bolivar to City of Hope to witness the detection of somatostatin in the bacterial colonies. Clustered in his lab, they instead saw with sinking hearts that the assay, sensitive though it was, detected not an iota of the hormone. The onlookers were devastated and Swanson particularly so. His horror at the shattering result was still palpable some twenty years later: "We were down at Art Riggs's lab at City of Hope looking for somatostatin [whispering theatrically] *and— nothing—came—out!*"[41] Riggs, characteristically understated, recalled being "not happy at all."[42] The next morning Swanson, a guest of Riggs, appeared deathly white and checked himself into a hospital emergency room. His pallid complexion and stomach upset turned out to be due to acute indigestion, accentuated by horror as he imagined his company, his career, and the money he had raised "going down the tubes."[43]

The scientists, accustomed to research setbacks and with less riding on the experimental outcome, quickly recovered. Surmising that bacterial enzymes had destroyed the tiny somatostatin protein as soon as it was made, they came up with a new experimental scheme. They would try to produce the hormone as a short tail on a much larger portion of the bacterial protein. The hybrid protein, they speculated, would be too large for the bacterial enzymes to degrade. In the final step, the scientists planned to chemically sever the somatostatin chain from the bacterial protein. If all went well, they would produce free-standing somatostatin. The strategy, based on scientific considerations, had an incidental political advantage: the hormone would remain safely nonfunctional until extracted from the bacteria and only then snipped free to become biologically active. There was little chance of bacteria containing engineered somatostatin running amok and risking infection, as some decriers of recombinant DNA research imagined. The authors of the subsequent publication, in a bow to recombinant DNA politics, pointed out the safety of the experiment in producing somatostatin as a "precursor" rather than in free and hypothetically dangerous form.[44]

The investigators put the new experimental plan into action and were soon getting results. In August 1977 the City of Hope team, without their UCSF colleagues, assembled to repeat the assay for somatostatin. This time the experiment worked like a charm. Itakura recalled the scene:

I think Heyneker sent some samples to the [Riggs] lab, and then Art was checking the immunoassay of somatostatin. Then we have about ten, maybe fifteen samples. Some samples are control[s], some ones are induction of the gene expression, some are not. Then we look at the printout of the radioimmune assay, and the printout show[s] clearly that the gene is expressed and somatostatin is there.[45]

The scientists were ecstatic. Riggs sent a sample to a Salk Institute scientist, who verified that it indeed had all the attributes of somatostatin. They had installed an artificial gene in bacteria and made a mammalian protein—and it worked like the real thing! Here was dramatic evidence that bacteria did not necessarily reject introduced genes and could manufacture a totally foreign protein using their own supposedly primitive cellular machinery. It was a first. As a delighted Boyer remarked, "We played a cruel trick on Mother Nature."[46]

Dazed by the sweeping implications, Riggs rushed off to a keep a date with his son for a baseball game at Dodger Stadium. He found he could not keep his mind on the action:

> I remember going there, but I don't remember anything about the game. I was thinking about all the incredible possibilities now that were likely to come to be. It's a spectacular setting. It's a beautiful stadium, and so it was not a bad place to be, to have your mind elsewhere, sort of contemplating miracles. . . . A whole world of opportunities suddenly became . . . much more likely.[47]

Fig. 11. UCSF and City of Hope scientists celebrating the successful somatostatin experiment, City of Hope, 1977. *Back row:* Art Riggs, Herb Boyer, Keiichi Itakura, and Roberto Crea. *Front row:* Lillian Shih, Herb Heyneker, Paco Bolivar, Eleanor Directo, and Tadaaki Hirose. (Photographer unknown; photograph courtesy of Roberto Crea.)

For Genentech, the achievement represented the critically important validation of its technology and consequent hope for a future. More broadly, the proof-of-concept experiment suggested a radically new procedure for making hormones and other useful products in bacteria. The experiment bridged the gap between the basic research preceding it—the Cohen-Boyer experiments and decades of earlier life science research—and the practical applications that followed. Heyneker recalled the episode with unalloyed pleasure:

> It was an *incredibly* exciting time. I felt on top of the world. . . . It opened up so many avenues [in applied research]; you could envision so many things you could do all of a sudden.[48]

Much remained to be worked out, at Genentech and elsewhere. But molecular biology had acquired a patently utilitarian dimension.

Somatostatin was by no means close to a marketable product and in fact would never become a commercial product for Genentech. Yet the research, contrary to predictions by reviewers of the Riggs-Itakura grant proposal, constituted a swift trajectory from the invention of recombinant DNA in 1973–74 to foreign protein production in bacteria in 1977. Basic and applied research had become proximate in a manner new to molecular biology. Furthermore, the proof-of-principle experiment had been conducted with notable economy. Boyer and Swanson had utilized existing scientific teams and academic laboratories because in them resided the rare and requisite body of knowledge and technical expertise. But as they and Perkins recognized, the strategy had the additional advantage of paying off handsomely in money and time saved. Swanson proudly announced Genentech's first research results at a meeting of private shareholders in April 1978:

> I am pleased to point out that the two year start-up of the company, including the completion of our first research goal, the production of the human hormone somatostatin, and the first commercial demonstration of our new technology, was accomplished for a total of $515,000. We plan to approach future growth in the same lean but effective manner.[49]

Management had cagily avoided investing precious start-up funds in leasing and equipping a laboratory facility and hiring a scientific team before anyone knew whether the technology had any chance of working as a productive industrial process. Cetus's Ron Cape later acknowledged as much:

> One of the mistakes [Cetus] made was not to realize the enormous leverage you get from using a university laboratory. . . . It is enormously cost effective. You're using labs and other goodies that are already there; you don't have to raise money and spend money to establish them.[50]

With patenting the somatostatin methodology as top priority, Tom Kiley came in to write and file applications for domestic and foreign patents. Without products to its name or even in the pipeline, Genentech's primary assets were its singular fund of scientific knowledge and technical know-how making up its all-important intellectual property. Securing timely legal ownership of its inventions was urgent, as Swanson and Kiley were all too aware. Furthermore, a patent portfolio—or, failing that, a collection of patent applications on file at the U.S. Patent Office— gave a young company without the standard metrics of products or profits a measure of credibility and a stronger negotiating position in the business and investment spheres. Kiley put it succinctly: "We thought we needed the protection of patents in order to justify investment in our company."[51]

Kiley had to move quickly to submit patent applications before Genentech's public announcement of the somatostatin results. Foreign patent law automatically invalidates patent applications if the research findings are previously published. He called a meeting of both research teams to discuss the proprietary claims. It was a learning experience on both sides. Before the group could get down to specifics, Kiley found he had to explain the patent system to scientists unfamiliar with its purpose and requirements. Unlike physical scientists and engineers, life scientists of the period trained and worked in an academic culture that placed scant emphasis on patents and intellectual property protection. The somatostatin research provided the impetus for Kiley to give his first tutorial on the importance to Genentech of patenting quickly, broadly, and well. It was a message he would have to repeat many times over to successive waves of newly hired scientists, many naive in intellectual property matters. Kiley for his part had to get up to speed in molecular biology. He read Watson's *Molecular Biology of the Gene* and impressed everyone with his quick grasp of the subject. The scientists gave Kiley a draft of the manuscript they intended to submit for publication and turned over their lab notebooks for review. They also needed to hurry to send off the manuscript for journal acceptance before Genentech's public announcement. Patent law in the person of Kiley was imprinting the sometimes leisurely pace of scientific publication with the urgency of proprietary protection.

Then a heated dispute over authorship broke out. Heyneker, claiming he had done much of the research, asserted that he rather than Itakura should have the honored place of first author. His native competitiveness fanned by the winner-take-all culture of UCSF molecular biology, he made a special trip to Los Angeles to discuss the matter with Itakura—to no avail. When the paper appeared in *Science* in December 1977, the names of the City of Hope chemists came first, with Itakura as lead author.[52] Kiley then had to make a critical decision—whom to designate as inventors on the patent applications. By the fall of 1975, if not earlier, Boyer had conceived of applying DNA synthesis and recombinant DNA in making hormones in bacteria, as his failed attempt to enlist the service of the German chemist attested. Nonetheless, Kiley determined that Boyer had not specified an explicit protocol for the somatostatin experiment. For determining inventorship, Kiley required evidence of a clear, detailed, and utilizable concept. He concluded that Riggs and Itakura had provided exactly that in their NIH grant application. Furthermore, the Boyer lab's use of recombinant DNA technology was not the novel invention; the Stanford-UC patent application of 1974 on the Cohen-Boyer procedure predated the somatostatin research by approximately three years.[53] Kiley named Riggs and Itakura as sole inventors on the patent applications. A UC patent administrator strenuously protested. According to Kiley, her view was that Boyer was a leader in the recombinant DNA field and at the very least should be named a co-inventor. Kiley's rejoinder was that he could find no evidence of "an inventive contribution on Boyer's part."[54] Not surprisingly, Riggs reinforced the point:

> Well, the key . . . was that [Itakura] and I developed the somatostatin plan. . . . We were the first to write down a concrete plan with sufficient detail to support a patent and to support the idea that an invention was made. Itakura and I were the ones that wrote and planned the NIH grant application.[55]

It was a disquieting moment for the UCSF scientists. They felt they had made essential contributions to the somatostatin research. More generally, it was an instance in which legal and scientific conventions were at cross-purposes. In patent law, those originating the concept for the invention are deemed inventors, and only their names appear on the patent application. Convention in scientific publication, on the other hand, is to acknowledge every contributor to an experiment by listing them all as authors. Patenting protocol tends to reduce the many to the few, while scientific publication protocol tends to expand the few to the many. In November 1977 Kiley filed four patent applications, making broad claims on the somatostatin research.[56] As it turned out, the U.S. Patent Office did not begin to review these or any other applications involving living organisms until the second half of 1980. All such filings were embargoed while the *Diamond v. Chakrabarty* case on the patentability of living things wended its way through the court system.

WIDER ISSUES

There were other clouds on the horizon. For recombinant DNA proponents, the most ominous development at mid-decade was the dozen or so bills and resolutions pending in Congress purporting to impose heavy restrictions on genetic engineering experiments. The biomedical research community, Cohen and Boyer prominently included, mounted an intense lobbying effort to persuade Congress not to impose legislative controls on scientifically significant research. But it was not only basic research that was at stake; future commercialization of the technology was in peril if Congress placed onerous restrictions on manufacturing recombinant products in quantity. As early as March 1977, before any company, aside from Genentech, had actually taken up recombinant DNA research, a federal committee had recommended legislation to extend the standards of the NIH guidelines to the private sector.[57] In this climate of opinion, Genentech's success in somatostatin counted as more than a mere technological achievement. Advocates of recombinant research seized upon it as a political argument against passing legislation threatening to nip scientific and industrial potential in the bud. An episode in the U.S. Senate highlights an early stage of politicization.

The occasion was a Senate subcommittee hearing on recombinant DNA in November 1977.[58] Paul Berg, the Stanford biochemist prominent in the political debate, and Philip Handler, president of the National Academy of Sciences, were scheduled for presentations.[59] Boyer had told them privately of Genentech's successful but not-yet-published production of somatostatin. All three recognized the achievement as a boon to their political cause—to help defuse congressional intent to pass crippling legislation on a field in which industrial prospects now seemed all the more likely. Handler and Berg decided on the spot to announce the somatostatin achievement to the subcommittee. Handler then proceeded to describe the engineering feat as a "scientific triumph of the first order." Berg, in similarly glowing terms, hailed it as "astonishing" evidence of the technology's promise in "putting us at the threshold of new forms of medicine, industry and agriculture."[60] The stamp of DNA politics on recombinant research was hard to miss. From now on, arguments warning of the risk of the United States failing to capitalize on the commercial applications of a field that it had inaugurated would thread prominently through the political debate. The incident also planted a seed of hope in life scientists that molecular biology, supported for decades by government money, might finally pay off in practical applications of use to the American public and in so doing help to keep the spigot of public research funds open and flowing.

The press picked up the Senate subcommittee story in a rush of articles, giving it the play that DNA proponents sought. Most quoted Handler's "triumph of the first order" statement and commented on the arrival of a breakthrough technology with an exciting industrial future. The *Washington Post*, for example, noted that it was not the achievement of somatostatin per se that was significant but rather "the practical promise" that man-made genes could make many useful products.[61] One trade publication addressed the biohazard debate head-on. The *Chemical and Engineering News* exclaimed in a headline, "Human Gene in *E. coli*: It Works!" and went on to call the somatostatin achievement "a vindication of the utility of recombinant DNA research which should further defuse a tiny group of scientific critics who claim that the technique is potentially dangerous to laboratory workers and to the public."[62] One could expect a pro-science publication to take a pro-science stance. But some general-interest articles also played up the industrial potential of recombinant DNA, making only passing reference to the safety issue.[63]

A *San Francisco Chronicle* science editor had a different reaction. He took indignant exception to the Senate subcommittee announcements, pronouncing them a violation of a long-standing code of practice in science journalism. Handler's and Berg's remarks, he protested in a letter to *Science*, were a violation of established protocol in science reportage and an overtly political maneuver:

> A double standard of scientific announcement seems to be operating here: The "orderly processes" of refereeing and publication remain in force for journalists and the public. But when the political process is operating in Congress—in this case, apparently, the spectre of political regulation for a new field of science—then the rules of science go by the board, and the public learns of a new scientific triumph via a Congressional hearing rather than through the pages of *Science* or the annual meetings of the American Society of Biological Chemists.[64]

The practice of orchestrating press conferences and other forms of publicity to announce scientific achievements before the research had been peer reviewed and published would soon escalate as universities threw convention to the wind, seeking to capitalize on the publicity and financial value of faculty discoveries, particularly regarding the cloning of important genes,

Months earlier Swanson and the City of Hope public relations department, eager to announce the somatostatin research results, had reluctantly postponed a press conference, deferring to the very concerns that worried the *Chronicle* editor. Riggs and Boyer, anxious to be seen to follow academic protocol, had argued to delay publicity until the work had been peer reviewed and accepted for publication.[65] On December 2, 1977, the somatostatin manuscript safely in press at *Science*, Swanson, the two teams of scientists, and City of Hope officials assembled at a Los Angeles hotel to announce the findings to the waiting press. The scientists were excited and on edge. For most, it was their first direct media experience. Boyer took things in hand. Remarking on the decades of federal funding poured into molecular biology with little of practical nature resulting, he noted the shift toward utility that the somatostatin research signaled: "The man in the street can now finally get a return on his investment in science."[66]

A return, he meant, measured in the purer, cheaper, and more plentiful pharmaceuticals that he believed recombinant DNA technology would produce.

The joint City of Hope and UCSF news release put it concretely, claiming a genetic engineering feat representing "the first demonstrated practical benefit from recombinant DNA technology." "Virtually identical techniques," the release went on, "could be used safely in bacteria to produce complex biological substances ranging from insulin and other hormones to the enzymes used in industrial fermentation."[67] Only in the last sentence did the release mention Genentech, noting simply that the somatostatin research "was funded by Genentech, Inc." One can only surmise the reason or reasons for the scant notice given Genentech. Hesitation in a roiling political climate to publicize the institutions' relationship to a company formed to commercialize recombinant DNA? A reluctance to share credit? Or another reason entirely?

In the subsequent media coverage, an exultant Boyer, eager to counter critics decrying recombinant DNA as a technology capable of engineering possibly perilous bacteria, described Genentech's pioneering use of synthetic DNA as a way to avoid any suspicion of danger:

> We've bypassed the potential hazards in recombinant DNA research. The gene is manufactured in a test tube. It's clean and has no contaminants. This [synthetic DNA and recombinant DNA technology] bridges the gap between chemistry and biology. These two disciplines are now married, and I think it's a marriage made in heaven.[68]

Paul Berg, a highly visible spokesman in the recombinant DNA controversy, ventured to the press that the use of chemically synthesized genes "can completely change the whole picture of estimating risks. You can design the gene so that even if something does go wrong and the bacteria [sic] finds its way to the human gut and survives, its product has little chance of doing any harm."[69] Genentech had chosen the synthetic DNA approach for scientific reasons; Berg noted a largely inadvertent risk-reduction approach in convenient conformity with his stance as science's statesman of responsibility in recombinant DNA research.

For Boyer, the flurry of publicity over the somatostatin research led to notoriety as well as accolade. In the steamy politics and polemics of the recombinant DNA political debate, he was a prominent and newsworthy focal point—on the one hand, lauded for a revolutionary invention, on the other, crucified for efforts to commercialize a controversial technology. For a scientist intent on having his research made practical and his company become a success, it was a boon to have the media play up Genentech's commercial prospects. But the publicity also spread word of Boyer's direct involvement in the company, up till then not broadly known.

Public knowledge of Boyer's corporate ties brought to the fore conflict-of-interest issues that in various forms would rankle academics and policymakers into the twenty-first century.[70] Put simply, what was the appropriate relationship of a professor to industry? How directly, if at all, should a faculty member involve him- or herself in a company? Where was the appropriate line between academic and commercial activities? How indeed could one distinguish basic and applied research, when in the case of somatostatin and numerous other research projects it was a seamless continuum? These questions racked university campuses as discoveries in molecular biology—and the professors making them—became increasing relevant to industry.

As the first in a long line of molecular scientists who would span academia and industry, Boyer felt the full and discomfiting weight of collegial censure. UCSF's campus newspaper reported that news of his ties to Genentech caused "an uproar in the Department of Biochemistry at UCSF since it appeared that Boyer was wearing two hats in order to benefit himself."[71] Two departmental members were particularly vocal in their profound disapproval. Such intimate corporate involvement, they asserted, represented a conflict of interest, if not a betrayal of academic values and traditions.[72] Professors of biomedicine, the ideology went, were to dedicate themselves to research, teaching, and public service; they were not to engage directly in for-profit companies. Biochemistry chairman Rutter let the uncomfortable issue more or less ride, but it roiled his department into the 1980s.[73] In 1981 he and a former student, fully attuned to the industrial prospects of their research on hepatitis B, would found the biotechnology company Chiron Corporation in nearby Emeryville.[74]

The outcry against the commercial trend was by no means confined to UCSF. For example, despite his promotion of the somatostatin work in the Senate a month earlier, Paul Berg told a reporter that the commercial involvement of professors was "just not to my taste. This isn't to criticize Herb particularly, but I just can't see it."[75] But in 1980 Berg himself stepped onto the corporate bandwagon. That year he, Arthur Kornberg, and a Stanford colleague cofounded DNAX, a private research institute for the commercial application of molecular and cellular biology. Berg's outlook had obviously evolved.[76]

News that Boyer had performed research in his UCSF lab under a Genentech contract added fuel to the gathering furor. Yet a professor's use of university facilities to perform contract research for industry had many precedents and, even with the waning of industry support with the post–World War II rise of federal research funding, remained fairly commonplace in American medical schools of the period.[77] Throughout the twentieth century, pharmaceutical companies had time and again contracted with academic researchers to perform the basic aspects of drug research and to run clinical trials. Historic examples are Eli Lilly's agreements on insulin with the University of Toronto in the early decades of the twentieth century and with Harvard and the University of Rochester on liver extracts to treat anemia.[78] The University of California likewise had a long history of industrial contracting, including a concurrent research agreement with Lilly on recombinant insulin.[79] In Boyer's case, the research contract with Genentech had the stamp of university approval, as verified by Rutter's and a university administrator's signatures on the contract document.[80] Then why, one might ask, was Boyer royally criticized and accused of conflict of interest?

Most likely it was his central role in the efforts to commercialize recombinant DNA technology, first through the Stanford-UC patent application and then through Genentech. Detractors, already disapproving of his position as inventor on a disputed patent application, greeted news of Boyer's corporate ties with renewed dismay and censure. The heart of his problem, as they saw it, was that as a full-time, tenured professor he was simultaneously and inappropriately cofounder, vice president, board member, adviser, and major stockholder of a private company—*Boyer's* company. These were not the usual arm's-length, part-time relationships with the business world that university faculties and administrators generally condoned. It was Boyer's direct, substantial, and ongoing association with Genentech, as well as his position as inventor on the suspect patent application, that academics found troubling in an era before such dual loyalties were commonplace and widely accepted. As his severest critics put it, he was "selling out to industry."

A low point was a UCSF faculty committee inquiry into the Genentech-sponsored somatostatin research transpiring in Boyer's lab. The resulting confidential UCSF committee report of 1979 noted the disruption and jealousy the contract research had evoked within the biochemistry department. It advised that "in the future it would be wise to refrain from making contracts in which work will be done by a university faculty member who also has a *major* financial interest in a concern, as this amounts to a contract between the person and himself, with the university's role only being incidental."[81] The report by no means ended the matter, at UCSF or elsewhere. In the future, research universities would expend countless hours on creating policy for ever-increasing numbers of faculty interactions with industry while steadily multiplying their own institutional ties with the corporate world.[82] J. Michael Bishop, a former colleague and 1989 Nobel laureate, later summed up Boyer's dilemma:

> Commercialization of biological discoveries was far from novel at the birth of Genentech: Big Pharma had been doing it for a long time. But for a member of the academic community to be so intimately involved, that was a sea change. No one had thought much about the rules for how this might be done. So there were repercussions, particularly among the faculty of UCSF—a hue and cry over potential conflicts of interest. It was a harrowing time for Herb Boyer.[83]

The opposition caught Boyer by surprise. Facing critics on several fronts—in the general public and, more disturbingly, within his home institution—he repeatedly argued that Genentech was a means to transform the invention of recombinant DNA into medical products useful to society. "I wanted to see," he told a science reporter in 1977, "that the technology gets transferred to private industry so that public benefits come out as soon as possible."[84] As he saw it, the somatostatin experiment was a test of that proposition and entirely appropriate for pursuit in his lab. Convinced that Genentech was set to contribute medical therapies for the public's well-being, he may in his enthusiasm have overlooked the possibility that others would, and indeed did, come to a different interpretation. His open promotion of Genentech among his UCSF colleagues, even suggesting that they purchase stock, appears to support this possibility. Looking back on this period, he commented:

It was very difficult for me. I had a lot of anxieties and bouts of depression associated with this. Here I thought I was doing something that was valuable to society, and doing something that would make a contribution, and then to have the accusations and criticisms, it was extremely difficult.[85]

As molecular biology's pioneering professor-entrepreneur, the first among his academic colleagues "to go commercial," Boyer was for some years a lightning rod for opponents of the industrial interests entering molecular biology.

While Boyer experienced a roller-coaster ride of professional and personal highs and lows, the pharmaceutical industry monitored recombinant DNA science and politics with a combination of fascination and skepticism. That skepticism began to fade somewhat with Genentech's making of somatostatin. Perhaps this radical and disruptive technology, corporate boards began to consider, could indeed be commercially productive. The title of a *Business Week* article on the somatostatin research, "A Commercial Debut for DNA Technology," reflected growing consensus on the technology's industrial relevance, especially in pharmaceutical manufacture.[86] A photograph accompanying the article reinforced the impression of a technology ripe for application. It pictured a grinning Boyer and Swanson standing before a blackboard diagram of the somatostatin experimental procedure. Boyer, with explosive curls somewhat tamed, was dressed in the suit, white shirt, and tie of a man of commerce. "The field is opening up rapidly," the article quoted him, "and we [at Genentech] have the flexibility to move." The message was clear: this new company with the high-tech name aimed for immediate impact on pharmaceutical production.

The pharmaceutical industry had gotten the message and begun to act. The *Business Week* piece went on to claim that as of late 1977 possibly as many as fifteen corporations were exploring the industrial applications of genetic engineering. Abbott Laboratories had reportedly begun work in the field, and Upjohn planned to inaugurate a recombinant DNA lab. Cetus had finally begun exploratory research in its new state-of-the-art recombinant DNA research facility. Hoffmann–La Roche had decided to hold back, awaiting more conclusive evidence that the strange technology could indeed produce marketable products. The flurry of corporate activity inevitably provoked criticism. Jeremy Rifkin, who would become a perennial burr in the saddle of biotechnology, had come out with an article in *Mother Jones* decrying the industrial trend and listing six major drug companies as engaged in or tooling up for recombinant DNA research.[87]

Genentech's somatostatin success had provoked a wave of corporate interest in genetic engineering. Yet the truth was, the hormone had no assured clinical application nor a clear and certain market. It was a convenient test substance and, as Swanson divined, not anything one could build a company upon. From the start, his bet had been on insulin, and he continued to hew to that view. With somatostatin achieved and the firm's technology preliminarily verified, he at last had the scientists behind him: insulin—*human* insulin—became the target in everyone's sight line.

4

Human Insulin:
Genentech Makes Its Mark

Human insulin has been produced at last by genetically engineered
bacteria in a California laboratory—an achievement that catapults
recombinant DNA technology into the major leagues of the drug
industry.
        Science News, September 16, 1978[1]

Swanson had impatiently endured the somatostatin project, with its
heart-stopping low point and exhilarating finish. It served handsomely
as a convincing proof-of-principle demonstration of Genentech's core
technology, reducing technical uncertainties and pointing toward a
sweeping landscape of industrial possibilities. After the somatostatin
success, the company could move on to a far more challenging project—
making human insulin. Swanson's urgent need, before the money ran
out, was to show that the company could make substitutes for the larger
and more complex proteins used in common medical practice. Human
insulin was that test. The project placed Genentech in direct conflict with
two elite academic teams, already well advanced in their quest for insulin. The ensuing cutthroat contest became all the more so for the promise
of concluding lucrative research agreements with a major pharmaceutical company. The competition for insulin was forbidding enough. But it
was a particular challenge for Genentech, which was little more than a
virtual company existing largely on paper except for an inconsequential
physical presence in Swanson's rented office in San Francisco. To launch
the ambitious insulin project, Swanson urgently needed lab space and a

scientific staff, and he needed them *fast!* At the beginning of 1978, his priorities were to locate and lease a facility, hire scientists for a trial-by-fire attempt to make human insulin, and attempt to raise yet more private money.

SEEKING CORPORATE CONTRACTS

Funding a start-up company was almost always an arduous job, all the more so one in a radically new biomedical field that financiers and corporate executives had no background for understanding and evaluating. Swanson, undaunted, concentrated his buoyant salesmanship on the task. Venture capital could support a start-up's creation and early development but as a rule could not provide long-term financing. A basic premise of venture investment was prompt monetary return to investors, either through a public stock offering or through acquisition. In the interim, a young company could turn to contracts with established companies to supplement the fits and starts of risk capital investment.

Early on Boyer had suggested that Genentech strike research and development agreements with pharmaceutical companies.[2] But it was not only access to their deep pockets he and Swanson wanted. To see a drug to market, Genentech would need the pharmaceutical industry's expertise in drug development, manufacture, and regulatory approval. Negotiating research contracts with established companies became yet another priority. Swanson, probably with Boyer and Perkins assisting, tried to interest Novo Industri, a Danish company dominating the European insulin market, in forming a partnership on human insulin. Novo went so far as to send delegates to Genentech early in 1978 but, questioning whether recombinant DNA could work as an industrial technology, decided against an alliance. Swanson also approached Hoechst, the German pharmaceutical and chemicals manufacturer. For similar reasons, it also turned its back on a partnership.[3] That left Eli Lilly and Company with an 80 percent share of the North American insulin market as Genentech's remaining best option for a contract.

The pharmaceutical giant, renowned since the 1920s for its market in pig and calf insulin, was the venerable patriarch of North American insulin production. By the mid-1970s, however, Lilly was seeking alternatives to extracting insulin from animal pancreases, a source not expected to keep up with the predicted expansion of the diabetic population. For good reason—to protect a flagship product—it kept a watchful eye on new methods for making the hormone. Irving Johnson, vice president of Lilly's research laboratories, identified, perhaps through publicity on somatostatin, recombinant DNA as a technology the drug maker should somehow acquire.[4] But well aware that academics tended to denigrate the quality and conditions of industrial research, he correctly discerned that even if Lilly's conservative upper management decided to create an internal research unit on recombinant insulin, the company would not succeed in hiring molecular biologists with the required skills in genetic engineering. Johnson then made it his vigorous aim to convince management that to remain competitive Lilly needed to form alliances with the academic teams pursuing research on recombinant insulin. "I felt my responsibility," he remarked, "was to make sure that Lilly was the first company to have human insulin."[5] In May 1976 Lilly sponsored a symposium on the genetic engineering of insulin. All the major players in the field attended, including Bill Rutter and Howard Goodman at UCSF and Walter Gilbert at Harvard.[6] The presentations gave Lilly, and Johnson in particular, a picture of current recombinant insulin research: the UCSF and Harvard teams appeared to be in the lead and clearly the ones to watch. By 1977 Johnson was periodically flying out to San Francisco to keep close tabs on the insulin research feverishly pursued in Rutter's and Goodman's laboratories. He appears not to have had a close connection with the Harvard team, perhaps because Gilbert was considering commercial possibilities of his own.[7]

PROCURING A FACILITY AND STAFF

Swanson meanwhile was pressed to find a facility for Genentech. The firm's operation from the Kleiner & Perkins suite in Menlo Park and then from an office in San Francisco's financial district had sufficed while the company was essentially Swanson, a part-time secretary, and a telephone. With Genentech's technology proven and the insulin project pending, it was high time to acquire lab space and hire scientists. In March 1978 Swanson completed a third round of private financing, providing $950,000 at $8 per share.[8] Genentech now had the means to move on to the next stage of corporate development. After considering various locations, Swanson and Perkins met with the mayor of South San Francisco, who encouraged them to locate in "The Industrial City," as block letters proclaimed on a freeway hillside. A few miles south of San Francisco, the city offered reasonable rent and a quick commute to UCSF and Stanford, active centers of the science, techniques, and workforce the company needed. South San Francisco, Swanson noted, was also close to an international airport and a short drive to Kleiner & Perkins and to other venture capital partnerships around Stanford. Silicon Valley—with its high-tech industries, support services, and entrepreneurial culture—lay just to the south.

Swanson and Perkins also weighed the local political situation. Unlike U.S. centers of unrest in San Francisco, Cambridge, Ann Arbor, and Berkeley, South San Francisco was not contemplating ordinances unfavorable to genetic engineering, and its citizens were unlikely to take to the streets in protest. Leery of Berkeley's history of political activism and the city council's tough ordinances regulating recombinant DNA research, Perkins observed that they chose South San Francisco in part because "it was not Berkeley. We perceived that Cetus would have a lot of trouble [with restrictive ordinances] in Berkeley, and they did."[9] In February 1978 Swanson leased a 10,000-square-foot section of an airfreight warehouse at 460 Point San Bruno Boulevard on the bay in South San Francisco.

Swanson, resolute in his vision of Genentech as a self-sufficient operation, had sought and found a site with potential for growth. He described it to private shareholders as "our new 'world headquarters,'" grandly predicting that it would become "a great campus someday."[10] Both claims were enormous leaps of faith—the portion of the warehouse Swanson had leased was an empty shell, without equipment or furniture—and, more to the point, without scientists. Swanson was undeterred and thinking big. Late in 1977, before he had leased space or hired a scientist, he had taken the premature step of employing a fermentation expert formerly at the pharmaceutical company Squibb & Sons, giving him the exalted title of vice president of manufacturing. There of course was nothing to manufacture. Swanson's intent was to signal to the outside world, if it cared to notice, that Genentech, Inc., was not a mere research boutique; it planned to make pharmaceuticals in short order. But without a single scientist hired or a functional lab in place, that was far from the case. As soon as the facility was in vague working order, Swanson or one of his scientists made a habit of posing for publicity shots in front of the gleaming valves and dials of an imposing biopharmaceutical fermenter. The

Fig. 12. Herb Heyneker regarding Genentech's first pharmaceutical fermenter, 1978. (Photographer unknown; photograph courtesy of Herbert L. Heyneker.)

purpose of course was to project the image of Genentech as an operating pharmaceutical manufacturer.

Boyer all along had tried to recruit scientists to the fledgling company. Riggs and Itakura and their labs remained under Genentech contract and were prepared to synthesize the DNA for the insulin project. But where was the labor force for the molecular biology component? An obvious hunting ground was Boyer's own department. His inside track to the Department of Biochemistry's personnel and resources was of inestimable value to Genentech, but at an immediate cost to himself in sour collegial relationships. In the company's early years, Boyer would serve as an essential bridge between academic and industrial domains, across which all manner of ideas, technical know-how, tangible materials, and manpower would migrate. His immediate plan was to seek out junior scientists, whom he knew from firsthand experience were the ones adept in the latest genetic engineering techniques. He also suspected that these young people at the outset of their careers were more likely than their securely tenured professors inured in academic tradition to consider taking a job in industry, and a high-risk one at that. Boyer approached several postdoctoral students about employment at Genentech. Heyneker, having demonstrated keen proficiency in the somatostatin experiment and abundant enthusiasm for practical application, was an obvious candidate. Sorely tempted as he was, the terms of his Dutch fellowship obligated him to join the University of Leiden faculty when the fellowship ended. In October 1977 Heyneker and his young family returned to Holland, expecting to put down permanent roots.

By that time Boyer's recruitment effort had grown more urgent. The previous May UCSF had mounted a gala press conference to announce the Rutter-Goodman team's success in cloning the rat gene for insulin.[11] It counted as a sensational achievement at a time when any gene cloning provoked great excitement and particularly one for a substance as well known as insulin. Axel Ullrich, a postdoc in Howard Goodman's lab, had employed a new and still troublesome procedure called complementary DNA cloning to reproduce the rat gene. As technically adept as he was highly ambitious, Ullrich used a newly discovered enzyme to make a DNA proxy of the natural gene and then cloned it in bacteria.[12] A torrent of publicity, including an article and photograph in *Time*, made much of the UCSF Department of Biochemistry's success as a major step toward the ultimate prize—production of human insulin.[13] Yet despite the fanfare,

Ullrich had not expressed rat insulin in the bacteria. Nonetheless, the Rutter-Goodman team, with the rat insulin gene cloning as a scientific trophy and the technical expertise backing it up, appeared to have a jump start on achieving human insulin and sealing a research and development contract with Eli Lilly or one of its competitors. Rutter recalled that after publication of the rat insulin research, "Every company involved in insulin manufacture came to see us. They wanted the clones."[14]

To Boyer and Swanson, it was all too worrisomely apparent that the UCSF team possessed a competitive method for making human insulin and might relegate Genentech to the sidelines, if not to immediate demise. Determined to acquire expertise in complementary DNA cloning for Genentech, Boyer invited Axel Ullrich and Peter Seeburg, the department's experts in the technique, to become consultants. The two Germans, both molecular geneticists, were friends as well as competitors in aiming to set molecular biology ablaze with their cloning prowess. Ullrich—husky, headstrong, and ambitious—was rightfully convinced of his competence and value. Seeburg, his lean face accented by a droopy mustache, was Ullrich's equal in technical proficiency and ambition. He was attempting to clone and express the gene for human growth hormone.[15] Boyer also courted the Australian John Shine, an expert in DNA sequencing and yet another of Goodman's stable of talented young scientists. When the three expressed interest, Swanson followed up with contracts specifying the consultant fee, amount of support for their UCSF research, and number of shares in Genentech.[16] All three signed consultant contracts in 1977.[17]

But Boyer did not limit his recruitment effort to the young. Eager for Genentech to benefit from the stature and insight of senior scientists, he tried without success to convince several professors to consult for or, in Stan Cohen's case, to join Genentech. Cohen declined on the grounds that he was already a scientific adviser at Cetus and that becoming a principal in a company was likely to jeopardize his lobbying activities against legislation restricting recombinant DNA research.[18] Boyer's overtures to Rutter and Goodman got somewhat further. He and Swanson resolved to secure them as consultants on complementary DNA cloning.[19] As Swanson summarized:

> Of the group of people that understood what was going on [in cloning research] at that point in time, [Rutter and Goodman] were the leaders.

I wanted to get the best advice we could. . . . There was this competitive technology, cDNA [complementary DNA] cloning technology. So I wanted to make sure that if it progressed, I had access to that technology and to the leaders. And Goodman and Rutter and Ullrich and Seeburg were on the forefront of that.[20]

In spring 1977 Swanson presented consultant agreements to Rutter and Goodman, which he had made pointedly contingent upon "delivery to Genentech of plasmids containing the entire gene coding for rat insulin."[21] He had placed the same demand on Ullrich, and doubtless also on Rutter, Seeburg, and Shine.[22] Swanson wanted the five scientists *and* their precious gene.

Perkins entered into the campaign for consultant expertise and for the biological material the UCSF professors controlled. He hosted a dinner at a high-end restaurant at which he did his persuasive best to cajole Rutter and Goodman into accepting the terms of the consultant agreement. He made no bones about knowing next to nothing about the ins and outs of molecular science, but he appreciated the value to Genentech of enlisting two prominent professors—Rutter, an academic powerhouse, and Goodman, a master of the latest genetic technologies. In the end, the effort to enlist all five UCSF biochemists came to naught. Rutter declined and Goodman and the three postdocs pulled out of their consultant agreements.[23] Negotiations broke off for several reasons, including failure to agree on the number of shares to be allotted. Doubtless also on Rutter's and Goodman's minds was the likelihood of concluding a research and development agreement with Lilly. Becoming Genentech consultants would surely diminish if not destroy the possibility for a UC-Lilly contract. Many years later Goodman put it succinctly, "We [Rutter and I] decided that a better partner [than Genentech] for the work we wanted to do was Eli Lilly."[24] Boyer's recruitment effort in his home department had reached a dry and dismal end.

But it was not only the UCSF biochemists who worried Boyer and Swanson. The formidable Wally Gilbert, directing Harvard's recombinant insulin effort, was a force to reckon with. He directed a talented group of cocky young researchers, all well aware of their standing in one of molecular biology's hottest labs. Soon to become a Nobel laureate (1980), Gilbert had already chalked up substantial contributions in gene regulation and rapid DNA sequencing. His lab used a complementary DNA cloning method, similar to that of the UCSF team, in its intense effort to trounce all comers in constructing, cloning, and expressing an insulin gene. But in the summer of 1976, politics had delivered the Gilbert team a serious blow. The Cambridge City Council and its firebrand mayor had issued a moratorium on all recombinant research within the city, incensed over its possible biohazards.[25] The decree brought Harvard's insulin research to a standstill. When the moratorium lifted in February 1977, Gilbert and his scientists were once more on course, racing to make up for lost time in their madcap attempt to express recombinant insulin and beat their UCSF competitors to the prize. To let off steam, the Harvard youths produced a run of newsletters taunting the West Coast team. One issue— titled "Gilbert Hustlers Outmuscle Boyer Cartel in Dual Meet: Coast Crew Crumbles as Gilbert's Gapes"—used the metaphor of a football game defeat to put the UCSF team in its place as a sorry loser to Harvard's elite.[26] But the barbed humor failed to disguise the ferocity of the east-west contest. At stake were scientific repute, corporate alliances, and the prestige of being first to clone a human gene for a celebrated medical substance.

Then, as 1977 wound down, Boyer and Swanson's hiring prospects took a turn for the better. Through Heyneker, Boyer and Swanson learned of Dennis Kleid, an organic chemist whose training in DNA synthesis and molecular biology seemed a perfect fit for Genentech's program. Fresh from postdoctoral fellowships at MIT and Harvard, Kleid in 1975 had joined Stanford Research Institute (SRI), a contract research organization in Silicon Valley, as director of its first DNA-based lab. A California native, the warmhearted Kleid was glad to settle in near the hotbeds of genetic engineering at Stanford and UCSF. He began to attend seminars at both universities and got to know Heyneker and others at UCSF. Late in 1977 Kleid received a call from Boyer and agreed to join him and Swanson for dinner at a trendy San Francisco restaurant. He learned that they were not interested in arranging contract research at SRI, as he had assumed, but instead wanted to recruit him to Genentech to work on an insulin project. Money to support the research would be no problem, Swanson assured him, obliquely alluding to a contract he expected to conclude with a pharmaceutical company. Caught off guard, Kleid hesitated. He had a responsible job at SRI. Why should he join a precarious venture without a laboratory or a scientist to its name?

Hearing of the offer, David Goeddel, Kleid's junior colleague at SRI, had none of his supervisor's doubts and hesitations. The strapping young Californian was a passionate rock climber and accustomed to taking risks, often deliberately seeking them out. Some years earlier, Goeddel had chosen the University of Colorado at Boulder for graduate work in biochemistry, more for its location in one of the world's great rock-climbing centers than for its scientific reputation. He joined an organic chemistry lab and tried his hand at DNA synthesis. But it was a course in molecular biology that caught his fancy. It was 1974, and word of the Cohen-Boyer method was trickling out. Perhaps, Goeddel thought, he could clone the synthetic DNA he and his lab partner Dan Yansura had struggled to make. As always, he was in a hurry. "I tried to get everything done as fast as I could," he recalled. "I wanted to graduate as fast as I could."[27] He yearned to return to California and the rock-climbing challenges of Yosemite and elsewhere in the Sierra Nevada.

Goeddel graduated in 1977 and got his wish. Willingly turning his back on academia, he accepted Kleid's invitation to join him at SRI, seduced by the opportunities that northern California offered to a technical climber. Kleid's description of what these climbs entailed throws light on Goeddel's winner-take-all mentality, whether applied to rock climbing or research:

> While Dave worked at SRI, he climbed El Capitan [a 3,000-foot vertical rock formation in Yosemite Valley], and that's a one-week thing. He explained to me how he does this: you have a rope, and you get a certain distance, and after about a day your rope doesn't reach the ground anymore, and you can't go back down. You have to go up. There's no way to go down. *You must complete the trip.* If you get nervous, your fingers sweat, you fall right off. So you have to be absolutely focused and confident and just climb that rock.[28]

For Goeddel, a job at Genentech offered easy access to the outdoor activities he craved. But he was also inexorably drawn to the challenge of pioneering applied research at the edge of do-ability—and the thrill of competing against Gilbert and his self-satisfied crew. What he saw in the fledgling company was an opportunity, demanding all his considerable technical and psychic resources, to develop frontier science for practical ends. Goeddel persuaded the still-wavering Kleid to get back in touch with Swanson and accept the job offer on both their accounts. Kleid relented and agreed to join Genentech as a senior research scientist, but on the proviso that Swanson would also hire Goeddel. In the end, he was

GENENTECH, INC.
460 POINT SAN BRUNO BOULEVARD
SOUTH SAN FRANCISCO, CALIFORNIA 94080
(415) 952-0123

EMPLOYEE'S PROPRIETARY INFORMATION
AND INVENTIONS AGREEMENT

Genentech, Inc. is dedicated to a policy of exerting a
dominant influence in molecular biology through continuing
technical superiority. The competitive success of this
policy depends to a large extent on the following two
factors:

1. The ability to capitalize on the creative
   talents of its employees, and

2. A free flow of information among its
   employees.

For this reason, all employees are requested to sign
the usual type of Agreement under which:

1. The Company is assured of title to inventions
   which relate to Company business, and

2. The Company is protected against unauthorized
   disclosure of proprietary information.

_____
ROBERT A. SWANSON, PRESIDENT

Fig. 13. First page of the 1978 Genentech employment agreement. (Copy courtesy of Roberto Crea.)

also seduced by the lure of doing cutting-edge research: "I went for the science, no question,"[29] he subsequently asserted. Swanson deferred to Boyer to size up twenty-six-year-old Goeddel's scientific qualifications. Boyer immediately recommended hiring him—a remarkably wise decision, it would turn out. Kleid and Goeddel then signed employment agreements, giving Genentech title to all inventions and protecting the company from unauthorized disclosure of proprietary information, a routine practice in industrial research labs.[30]

Late in 1977, perhaps through the somatostatin publicity, Lilly's Irving Johnson learned of a new contender—an unprepossessing start-up with a grand plan to launch research on human insulin.[31] By February 1978 Johnson and Swanson were in touch, Swanson pressing hard for a research and development contract with the venerable drug firm. It surely took all of his considerable promotional talent to represent Genentech, without a scientist in place, as a credible enterprise, able to outpace its prestigious academic competitors and produce the vaunted hormone. Johnson convinced his dubious superiors at Lilly to keep all options open and reach into their deep corporate pockets to fund the improbable effort that the infant company aimed to launch that spring. In June Lilly and Genentech reached a preliminary understanding in which Lilly agreed to support Genentech's insulin effort at $50,000 a month.[32] Lilly had signed a similar agreement with the University of California the previous March, but on growth hormone as well as insulin. Lilly's contracts with two competing institutions suggested how badly it wanted a human insulin product. Although Lilly was intent on covering all bases through modest research support, it was of no mind to sign a formal long-term R&D agreement with any institution without strong evidence of its capacity to make human insulin.

The City of Hope chemists, still under Genentech contract, had for some months been hard at work on the synthesis of DNA fragments coding for insulin, the second hormone specified in the medical center's contract with Genentech. As Riggs remembered it, one night they celebrated somatostatin, the next day they started on insulin.[33] Making the DNA for human insulin was no easy endeavor. The molecule is considerably larger and more complex than somatostatin's—as mentioned, fifty-one amino acids compared to somatostatin's fourteen. It consists of two amino acid chains, the so-called A and B chains, which required the chemists to synthesize two DNA sequences, one for each chain. Itakura and Crea spent several days designing the chemistry. As in the somatostatin research, they were not aiming to make an exact copy of the natural gene. In fact, the complete sequence of the human insulin gene was unknown. Instead, they worked back from the amino acids composing the insulin molecule and, guided by the genetic code, selected DNA sequences for chemical synthesis that they believed compatible with bacterial cell machinery.[34]

It was a strategy designed for a utilitarian purpose—to force bacteria to produce a human protein.

Crea led the DNA chemistry, with several postdocs and technicians assisting. The team endured long days of exposure to the noxious chemicals and toxic fumes of DNA synthesis to create and purify the fragments for the DNA sequences coding for the two insulin chains. Crea recalled the feverish intensity:

> There was so much excitement in the lab that even occasional episodes of solvent spills did not bother us. We were cranking, literally, molecule after molecule in a frenetic race to get there first and fast. We didn't spare anything. We put our heads down, and we created a beautiful episode of efficiency and productivity.[35]

He was living up to his name: Crea in Italian means "he creates." The team finished the chemistry within six months, remarkable speed considering the arduous and noxious labor required before the advent of gene-synthesis machines. The work then shifted to Genentech.

In mid-March 1978 Goeddel arrived to begin work at Genentech. Kleid came in April. The firm had acquired its first resident scientists. They found the DNA fragments from City of Hope waiting for assembly. But what formidable obstacles they faced. Late runners in the contest for insulin, Genentech's scientists, north and south, lagged well behind the accomplished teams at UCSF and Harvard, already considerably advanced in a contest to out-compete each other and capture the cash and kudos for making human insulin.[36] Genentech's laboratory at one end of the warehouse was completely empty, a hollow shell. "I'm talking the four walls, the ceiling, and the floor,"[37] Kleid recalled. He and Goeddel had to find alternate lab space, and quickly. Riggs, coming to their rescue, arranged for their use of a closet-size lab at City of Hope. The two immediately headed south, determined to make up for lost time. Within weeks they had the A-chain genetic sequence assembled and inserted into plasmids for cloning. The B chain was another matter. Goeddel labored obsessively to clone it but made little progress. He paused only to eat and drop into exhausted sleep when Kleid took over the shift. Itakura, no slouch himself, observed that Goeddel was a "crazy hard worker," toiling as much as twenty hours a day and applying his do-or-die tenacity.[38] After a number of frustrating attempts, they discovered that the cloning problem lay in a mistake in the B-chain sequence.

Late in May 1978 Herb Heyneker turned up unexpectedly on a visit from Holland. Boyer and Swanson's standing offer to join Genentech had finally paid off. Wining and dining him in Amsterdam late in 1977, they had pressed him to become Genentech's first director of molecular biology. This time Heyneker had jumped at the offer. He had found himself entrapped in the Byzantine coils of DNA politics, all but unable to conduct recombinant DNA research in Holland. A Dutch government advisory cautioned against experiments in the field, creating a situation close to a research moratorium. On no account was Heyneker, competitive to the bone, prepared to lose his pioneering position in genetic engineering, as he explained:

> It really became clear to me that I didn't want to stay in the Netherlands under those conditions. I was at the forefront of this technology in the United States, and to stop this line of research, knowing that my colleagues and other researchers would continue and move on from where I left off, was difficult to swallow. . . . It was something which I had at my fingers[tips]; I had done it for two years; I knew exactly what to do.[39]

He agreed to join Genentech in September 1978 at an annual salary of $40,000, after his wife had delivered their third child.

Heyneker's unexpected visit to California in May was to attend a conference and buy a house in the Bay Area. Instead, he learned from Goeddel of the problem in cloning the B-chain sequence. A more experienced cloner, Heyneker sidelined his house hunting and made a beeline to City of Hope. He and Goeddel, their young male energy peaking at the thrill of the chase, urged each other on, intent on surmounting the problem and showing their mettle as ace cloners. Failure for them was out of the question. Goeddel later remarked on his many subsequent sprints to the finish line: "It was not going to pay to come in second. You either came in first or you might as well be last."[40] He and Heyneker fell into a spontaneous rhythm of groundbreaking research done under exceptional pressure. Heyneker recalled:

> We worked so well together. [Dave] understood exactly what I wanted to do, and I understood quite well what he wanted to do. So we took turns

sleeping, to speed up things. When we had to [radioactively] label the DNA fragments, we were so much in a hurry that we exposed an X-ray film only for as short a time as possible and looked at an angle at the film so that we better could see a slightly darker position—everything to be as fast as possible. We had a bunch of [B-chain clones] within a week or five days. And we took them home [to the Bay Area], they were sequenced, and we were successful in getting the right B chain. . . . The experience was fantastic, very exciting—how efficient can you be![41]

Genentech's molecular biology research was no sooner up and going than it came to a grinding halt. Rumor was that Wally Gilbert and his Harvard team had produced proinsulin, an inactive precursor of insulin.[42] Using a complementary DNA approach, they had done what the UCSF team a year earlier had not accomplished: they had produced a preliminary form of recombinant insulin in bacteria. But it turned out on further information to be a precursor of *rat* insulin—not *human* insulin, the prize of all prizes. The West Coast scientists sighed with relief. They had panicked on false news that the Harvard group had made human insulin. Shortly after announcing the result, Gilbert struck a deal with Biogen, a genetic engineering company he and a number of leading European molecular biologists and venture capitalists had founded in Switzerland early in 1978.[43] With Gilbert's insulin research in full gear, Biogen chose human insulin as one of its prime targets and began to underwrite his research at Harvard. Lilly, anxious to appropriate any and all promising methods for insulin production, tried and failed to license Gilbert's insulin technology. Not surprisingly, the license went instead to Biogen.

In June Genentech's third scientist signed on in the modest person of Daniel Yansura. Goeddel had relentlessly pestered his former University of Colorado lab partner to join the firm and work with him on insulin. Yes, Goeddel admitted, there was a chance for failure—of the research, of the funding, of the entire venture. But the science itself was irresistible. It was an opportunity, Goeddel insisted, to take their earlier work a significant step forward, to actual production of insulin. Yansura was hooked: "That I thought was an incredible challenge," he explained, the opportunity still bright in his mind years later.[44] He was young—twenty-seven—unattached, and, like Goeddel, blithely unfazed at leaving academia and joining a highly speculative enterprise. Additional perks for the Detroit native were the California lifestyle and a higher salary—

$18,000, compared to the $12,000 he earned in Colorado. Swanson made clear that Genentech scientists, the cream of his budding enterprise, would be free of the endless grant writing and fund-raising that burdened academic life. "This is a science-driven company," he told recruits. "Don't worry about money. Anything you need, you've got."[45] But for Yansura, what was uppermost was the prospect of doing pioneering research to make useful products. He accepted the position.

Swanson offered Yansura, as he had done for Kleid and Goeddel, a chance to buy low-cost stock and acquire part ownership in Genentech. The custom of offering employees stock or stock options had been current for decades in high-tech Silicon Valley companies.[46] The practice was anything but commonplace in Big Pharma, where as a rule only top management held stock and/or stock options. The thinking behind the practice at entrepreneurial companies was that employees would remain loyal and work harder if they owned a piece of the company. However, in the 1970s academics like Yansura were often oblivious to the value of stock ownership. For Yansura, getting his first taste of corporate procedure, Swanson's offer to purchase cheap shares fell flat. Baffled, he hesitated overnight before deciding to invest $300.[47]

Yansura joined Kleid and Goeddel in the warehouse lab, by midsummer 1978 marginally serviceable. The three labored under the delusion that Swanson had wrangled a contract with Lilly.[48] If any one of them had the slightest doubt that they were entrenched in a hell-bent race to achieve human insulin, Swanson was there constantly to remind them, pointedly dropping Gilbert's name to egg them on. The three—eager to prove the technology, the company, and themselves—did not need Swanson's goading; they were already obsessively focused on making the hormone. Yansura recalled:

> Everybody at the time was fairly numb, and we worked long hours. . . . My wife, Patricia, reminds me that it was very long, that she spent most evenings during the week by herself, and that I would come in around nine or ten. . . . I would eat, sleep, and get going the next morning. Five days plus part of Saturday were expected. Everybody else was working hard so you just fell in.[49]

Motivation came from the young men's singleness of purpose, of together moving an exciting technology to the ultimate in useful productivity, and the camaraderie of the like-minded drive for success. The immediate motivator, however, was the contest with the Gilbert and Rutter-Goodman teams, which for the moment were either dismissive or largely unaware of the newest entry into the human insulin contest. For Genentech folk, however, the competition was all too clear. "We knew we were in a race," Yansura recalled, "and we knew we would have to be first to survive as a company."[50]

That summer of 1978, Goeddel, with Kleid and Yansura assisting, set about to clone and express the two DNA sequences coding for the A and B chains. Using the somatostatin experiment as a rough model, they spliced each A- and B-chain sequence into separate plasmids and transferred the hybrid plasmid into bacteria for cloning. As with somatostatin, the bacteria did not produce the chains free and clear. Rather, they were fused to a much larger bacterial enzyme as minute tails of insulin A or B chain. Goeddel then harvested the chains by chemically clipping them free. With no protein chemist to help them out, the three struggled to fish out the A- and B-chain proteins from the welter of bacterial material. Fortunately, they could call on Crea at City of Hope, who had the necessary instrument to purify the material. Then Goeddel succeeded in a marathon onslaught to isolate the insulin chains. The tenacity and endurance he displayed in scaling a Sierra rock face served him equally well in science.

What remained for Riggs and his lab to accomplish was the challenge of uniting the two chains to form an intact insulin molecule. He had meticulously researched the scientific literature and found a chain-joining procedure that he passed on to his lab.[51] July passed worrisomely with no positive results. Finally, a frantic Swanson, with Perkins breathing down his neck, ordered Goeddel down to City of Hope and forbade him from returning till he had assembled the two chains into an insulin molecule. Someone discovered that bacterial proteins had contaminated the B chain and obstructed the joining procedure. It then fell to Crea to toil another spate of exhausting hours to remove the impurities and then hand the material over to Goeddel. Ignoring hunger and sleep, Goeddel worked nonstop around the clock. Finally, in the small, lonely hours of August 21, he succeeded in reconstituting the two chains into the insulin molecule. Genentech had made insulin, *human* insulin.[52] It was a golden moment. Two teams of unknowns supported by an obscure company had come from behind, encountered problems at several turns, and

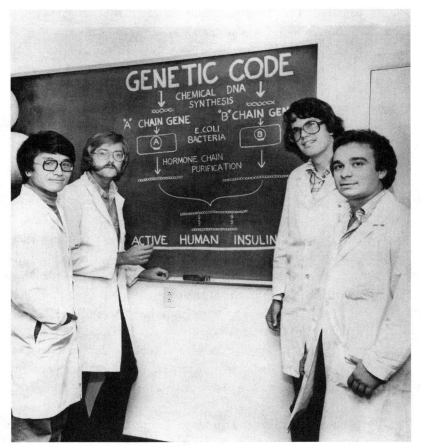

Fig. 14. Keiichi Itakura, Art Riggs, Dave Goeddel, and Roberto Crea at a blackboard with a diagram of the human insulin experimental scheme, Riggs's City of Hope office, August/September 1978. (Photographer unknown; photograph courtesy of Arthur D. Riggs.)

managed to out-compete two elite academic teams in making a human form of a celebrated hormone. For the youthful crew, that was supreme exhilaration.

Scientific ingenuity, technical proficiency, and relentless drive defined Genentech's triumph over rivals. Yet political circumstance and government policy also figured in the company's success. Because the original NIH guidelines applied only to recombinant experiments funded by the federal government, Genentech's privately funded research was technically exempt. Furthermore, the 1976 guidelines concerned natural and complementary DNA and contained no explicit reference to chemically synthesized DNA. The City of Hope chemists could therefore perform the gene synthesis work under ordinary lab conditions. As it turned out, the molecular biologists could also conduct much of their research with only the usual precautions. As Kleid explained: "99% of the [insulin] work was done basically with chemicals: synthetic DNA, plasmid DNA of *E. coli* extracts. So we didn't need to do much actual work in a closed-up [physically and biologically secure] laboratory."[53] Although Boyer and Riggs had adopted the DNA synthetic approach to gene construction for scientific and technical reasons—an approach they had adopted *before* publication of the NIH guidelines in July 1976—their use of synthetic DNA turned out to have important competitive as well as technical and political advantages: Genentech was not burdened by the crippling and costly early guideline restrictions on experiments involving natural human genetic material.

The UCSF and Harvard insulin researchers were not so fortunate. Because their approach used human genetic material, they were subjected to the full weight of the guidelines and their mandated safeguards. The most significant roadblock was the requirement to conduct their experiments at the highest level of biological containment, conditions available only in a handful of biological warfare labs. In the United States, one such facility existed at Fort Detrick, Maryland, and another at the navy's biological lab in Alameda, California. Neither UCSF nor Harvard could obtain access to either lab. In mid-1978 members of the two teams therefore went abroad, to countries in which the guidelines were less stringent than in the United States. Axel Ullrich went to Strasbourg, France, to work in a lab that Eli Lilly had hastily adapted to French safety standards for human DNA research.[54] There Ullrich isolated and cloned the human proinsulin gene but did not achieve expression. Gilbert and members of his Harvard team gained permission from the British military to use, for a scant one month, England's high-containment biological warfare facility at Porton Down.[55] On top of the extreme inconvenience of performing experiments in the sterile regalia and awkward conditions of high-level biosafety conditions, Gilbert's team returned home, greatly frustrated, having failed to obtain clones of the human insulin gene. They had mistakenly re-cloned the rat insulin DNA, rat material having contaminated their preparations in transit. Kiley recalled the situation with customary flair and hyperbole:

> On the very day when we were announcing success in insulin, [Gilbert] was, as he had for many days past, trudging through an airlock, dipping

his shoes in formaldehyde on his way into the chamber in which he was obliged to conduct his experiments. While out at Genentech we were simply synthesizing DNA and throwing it into bacteria, none of which even required compliance with the NIH guidelines.[56]

THE ELI LILLY CONTRACT

For Swanson, the insulin success meant immediate resolution of the interminable negotiation with Lilly. Genentech had accomplished what Johnson and his superiors had all along required: production of human insulin in material and measurable form. That done, Lilly rushed after months of dawdling to secure a research and development agreement, fearing that Genentech might license a competitor for exclusive use of its technology. Lilly's executive vice president, Cornelius Pettinga, and a patent attorney immediately boarded a corporate jet and flew to Los Angeles, at long last prepared to sign an agreement. Swanson, Boyer, Kiley, and Genentech's new director of finance, Fred Middleton, greeted them at Kiley's law office. It was a difficult and delicate process for the Californians, determined to stand up to the powerful multinational and protect Genentech's technology, its vital asset. On August 25, 1978—four days after Goeddel's insulin chain-joining feat—the two parties signed a multimillion-dollar, twenty-year research and development agreement. For an upfront licensing fee of $500,000, Lilly got what it wanted: exclusive worldwide rights to manufacture and market human insulin using Genentech's technology. Genentech was to receive 6 percent royalties and City of Hope 2 percent royalties on product sales.[57] For a firm barely off the ground, the agreement provided an immediate supply of much-needed cash, a likely income stream through benchmark payments, and an alliance with a prestigious pharmaceutical company with the know-how and worldwide repute in manufacturing and marketing insulin. In the eyes of the pharmaceutical industry, Genentech was now on the map.

But Lilly had done more than contract with Genentech. Unbeknownst to Genentech, the pharmaceutical giant had previously sealed an agreement with the University of California. On August 17, 1978—eight days before the Genentech-Lilly contract signing—Lilly and UC concluded a $1.3 million, five-year agreement on the complementary DNA cloning and expression of human insulin and human growth hormone. Lilly had first option to license the technology.[58] Years later Boyer commented on Lilly's simultaneous contracts with two competing institutions:

> You know, if you've got the money and you can afford it, cover all your bets. . . . I think Lilly's approach was, well, we'll give these [Genentech] guys some money, and if they can do it, good. It's not a big hit for our budget. If it turns out to be successful, we've got ourselves covered. What Lilly would be concerned about was if we could have demonstrated that it was possible to make insulin this way, they would have a major competitor.[59]

For Swanson, doing his dogged best to protect the interests of his fledgling company and above all retain control of its core technology, the negotiations with the pharmaceutical giant were harrowing. A youthful-looking thirty-one-year-old facing pharmaceutical executives far senior in age and experience, his worst fear was that Lilly would appropriate Genentech's technology, its crown jewel, and apply it in its own projects. Yet Swanson also fully recognized that without Lilly's financial resources and manufacturing and marketing acumen, the human insulin experiment was likely to remain little more than a basic-science demonstration. In that case, Genentech's future appeared dismally bleak. "I was in this negotiation," Swanson later remarked, "trying to figure out how a tiny company like ourselves could protect the technology, because obviously you're going to give it to someone else to make the product, but they could steal it. . . . People thought I was a little paranoid about it."[60] He and Kiley managed to negotiate a contractual condition limiting Lilly's use of Genentech's engineered bacteria to the manufacture of recombinant insulin alone. The technology itself would remain Genentech's property, or so they expected. As it turned out, the contract, and that clause in particular, became a basis for prolonged litigation. In 1990 the courts awarded Genentech over $150 million in a decision determining that Lilly had violated the 1978 contract by using a component of Genentech's insulin technology in making its own human growth hormone product.[61]

The contract stipulated that Genentech was to develop and provide to Lilly bacteria producing human insulin of a specified quantity and purity. To ensure that Lilly got what it wanted without wasting money on unproductive research, it specified a series of research benchmarks that Genentech was to meet by specific dates. Lilly would provide periodic milestone (progress) payments—but only if Genentech reached the benchmarks on time. If it failed to do so, Lilly had the option to terminate the agreement. The system gave Lilly control over the extent to which it sunk money into a commercially uncertain project and outlined an orderly, time-dependent progression toward specific production goals. Genentech in turn received crucial financing without diluting equity by selling stock. Johnson remembers the benchmarks as straightforward and a standard part of corporate research agreements.[62] For the Genentech researchers, saddled with rigorous deadlines and quotas, they were a rude awakening. Only after the deal with Lilly had been signed and sealed did they learn what the agreement committed them to achieving. Kleid remembered his distress:

> [The negotiators] had a very rosy outlook about what was going to happen. When I looked at this [agreement], they were not talking about a rosy outlook, they were talking about my future. They had signed something that said these people [Lilly] were going to be able to walk off with this bug [if] we couldn't make a certain [amount of insulin] by a certain date, [an amount] that was sixty times away from where we were. Sixty times! . . . I was very upset. I said, "Wow, this is impossible." Bob [Swanson] said, "Dennis, this is possible. What are you talking about?! It's not impossible. I don't want to hear that word impossible. Just tell me what you need to accomplish it."[63]

The three scientists were excruciatingly aware, although Swanson did not hesitate to remind them, that they bore Genentech's fate on their shoulders as well as that of their own salaries and job security. "Each time we reached a benchmark," Yansura remarked, "we would get more money [from Lilly]. Of course, that was our salaries."[64] Failure to reach the benchmarks on time could mean the firm would go the way of many an entrepreneurial start-up, forced to fold after burning up capital without achieving production goals. For the scientists, swelled to five with the arrival of Heyneker and Crea in September 1978, that searing possibility and the pressure of producing results to fit an industrial timetable was a sobering comedown from the exhilarating insulin press conference and flurry of media attention.

Looking back on the agreement after many years, participants differ on which side negotiated a tougher deal. Perkins believed that the

8 percent royalty rate was unusually high, at a time when royalties on pharmaceutical products were along the lines of 3 or 4 percent.[65] Johnson concurred: "It was kind of an exorbitant royalty, but we agreed anyway— Lilly was anxious to be first [with human insulin]."[66] On the other hand, Middleton, Genentech's financial officer, felt that Lilly did extraordinarily well:

> [Lilly] got such an unbelievably good deal on it, in retrospect. They paid three million dollars in milestones and a high single-digit royalty (8 percent). They made billions of dollars on the deal. I remember Neil Pettinga was saying [during the negotiations], "Oh, you [Genentech] guys are such tough negotiators." He just complained and complained. "Lilly never paid so much for anything, blah, blah, blah." He was so upset at the terms, he couldn't enjoy the toast. Swanson had gotten some champagne. He told me the next day it was a bad omen . . . for the relationship with Lilly because the champagne was flat when they opened it.[67]

In hindsight, the contract represented more than a legal seal on a business deal. It signaled the presence of a new organizational arrangement in the pharmaceutical industry—the big company–small company alliance. Genentech and future small research firms like it would function as entrepreneurial sites in which biology-based technologies were developed for large-scale manufacturing. As intermediaries spanning the university-industry divide, scientists at entrepreneurial start-ups would perform research generally too utilitarian for a university environment and too early stage for the average pharmaceutical company. As Swanson proclaimed in 1979: "We have bridged the academic and industrial worlds and forged a network designed to help us maintain our demonstrated position of technical leadership among academic and commercial groups that are pursuing applications of molecular biology."[68] Any putative distinction in genetic engineering between basic academic research and applied corporate research was fast evaporating. The linkages would grow in complexity and extent as the field advanced, making commercial biotechnology the most intricately intertwined with university research of any industrial sector.[69] The big company–small company template that Genentech and Lilly promulgated in molecular biology would become a prominent organizational form in a coming biotechnology industry.[70]

Swanson, hell-bent on establishing Genentech's priority in human in-sulin and advertising the Lilly contract, lost no time in scheduling a press conference. It was to be held at City of Hope since Genentech's headquarters, still being outfitted, were in no state to handle the media. On the day of the press conference, September 6, 1978, the San Francisco contingent—Swanson, Kleid, Goeddel, and Yansura (Boyer was abroad), dressed up for the occasion in suits and ties—flew down to Southern Cal-ifornia. The Genentech and City of Hope scientists, arranged on a stage in separate rows, faced an audience of thirty or forty, including Lilly repre-sentatives, television network personnel, and newspaper reporters. Flus-tered by a battery of cameras and lights, Riggs was so nervous that after-ward he could not recall much about the occasion. Goeddel was likewise overawed. For the scientists, the media limelight was a new and thrilling but intimidating experience.

The pivotal moment was the announcement that Genentech had made human insulin and signed an agreement with Eli Lilly. City of Hope's diabetes expert then reeled off a list of human insulin's supposed thera-peutic advantages and launched into a description of the research. The Genentech contingent was annoyed—a stranger unassociated with the research was explaining *their* research. Swanson elbowed Goeddel to take over. Overcome with stage fright, the intrepid rock climber froze and re-fused. Kleid rose in his stead and began addressing questions from the audience. Trying to finesse a query about the safety of recombinant in-sulin, he claimed that Genentech had not actually made insulin; it had made a fusion protein. Only in clipping off the superfluous bacterial protein, he maintained, had the scientists actually made the hormone. It was a convoluted point meant to allay the questioner's concern over possible biohazards. Swanson, determined to capitalize on the insulin achievement and avoid ethical issues at all costs, was "so furious [over losing control of the presentation] you could see him steaming," Goed-del recalled.[71] Swanson hurried to the podium. At that instant, the room plunged into darkness. The drain of the media lights had tripped an elec-trical circuit breaker. The press conference came to an abrupt and thank-ful end. But Swanson had gotten the publicity he sought—a platform for broadcasting Genentech's achievement and provoking investor inter-est. "He was very good at that," Goeddel reminisced.[72] The San Francisco group flew home to find they had made the front page of the *San Francisco Examiner.* A banner headline proclaimed, "New Insulin for Diabetics: Bay Area Labs Lead the Way." Swanson later ordered paperweights with replicas of the page.[73]

But not everyone was pleased. Bystanders pointed out that the insulin press conference had preceded peer review of the research itself. The two insulin papers were not communicated to the *Proceedings of the National Academy of Science* until October 2 and 3, almost a month after the public announcement. One critic, aware of UCSF's similar transgression in announcing the rat insulin success, bitingly labeled the trend "gene cloning by press conference."[74] Swanson, single-mindedly promoting his company, was considerably more interested in Genentech capturing media and investor attention than in conforming to the niceties of academic protocol. Riggs, however, had again been uneasy. At his behest, Itakura in a hasty bow to scholarly convention had presented the insulin work at UCLA the day before the press conference.

Pressed by the rapidly approaching news conference, Kiley had hurriedly expanded the somatostatin patent applications to include claims on the insulin findings, filing them only days before the media event.[75] The rapid pace of gene-cloning research and the perceived need for intellectual property protection would soon make the frantic dash to file patent applications before public presentations a common feature of the new science, whether in industry or academia. Genentech's scientists, steeped in academic tradition, assumed all along they would publish the insulin research findings in scholarly journals. Swanson, they found, was resistant. Cultured in business practice, his instinct was to keep experimental findings secret to protect the company's intellectual property from possibly being stolen and copied by competitors. The scientists were dismayed. They wanted their research findings published in time-honored academic tradition as contributions to the open scientific literature and a primary means to achieve individual scientific credit and professional status.

Boyer stepped in and settled the dispute: Genentech scientists would publish; in fact, they were to be *encouraged* to publish. He recalled:

> I insisted that we have the scientists publish their research in journals. Any proprietary information would have to be covered by patents. I felt this was extremely important for attracting the outstanding young

scientists in the community that were interested in doing research in an industrial setting.

I also wanted to bring in scientists that were outstanding and have them have an opportunity to establish their own reputation, get their own recognition. So we tried to set up an atmosphere which would take the best from industry and the best from the academic community, and put them together.[76]

Boyer's decision was in line with his academic training and professional orientation and what he knew any university scientist Genentech employed would expect and demand. It was also farsighted. Publication in peer-reviewed journals was a prime means for the firm to achieve priority of discovery and to display the high-quality and pathbreaking nature of its research. For a business with exceptionally close ties with university science and scientists, establishing credibility and a certain commonality of values was especially important in an era in which academics tended to ignore or even denigrate industrial research.[77]

But Boyer's mandate for open publication came with a hard-and-fast requirement. In a significant concession to corporate norms, Kiley would first file patent applications, if warranted, and review manuscripts before they went to press for proprietary information and confidential disclosures. Although charged with protecting Genentech's intellectual property, Kiley came to appreciate Boyer's decision as the years rolled by:

One of Herb Boyer's great contributions was insisting that Genentech publish its work. It helped us attract scientists who crave the peer recognition publication brings. It acted as quality control, that being one of the great points of refereed journals. If you can pass muster with the referees and get published in a reputable journal, you're doing good science. It helped to validate the company in the eyes of potential customers. If you will, it enhanced our celebrity. And it greatly aided us in recruiting the best and the brightest from academic centers, where traditionally they'd been chary of industry because of the perception, not inaccurate, that in industry and particularly in the pharmaceutical industry, trade secrecy trumped publication.[78]

Genentech was not unique in granting its employees publication privileges. Such rights were more or less common in high-tech companies in Silicon Valley and some research-based firms elsewhere. An open-publication policy traced back in some cases to the early twentieth century, to the research laboratories of corporations such as General Electric, AT&T, and even the relatively secretive DuPont, all of which recognized publication as a means to attract, keep, and motivate their scientists.[79] Yet in the pharmaceutical industry, intellectual property concerns and trade secrecy tended to dominate drug research, with company scientists by and large reticent to reveal research details to outsiders. A few drug companies were known to delay publication for years after patents were filed, to the detriment of staff morale and the firm's contribution to the shared scientific commons.

Swanson soon came to appreciate Boyer's publication policy, not because he had an impulse to instill academic values at Genentech or a driving wish to contribute to the progress of science. Rather, he learned over time that the policy paid off in the high-caliber scientists it helped to attract and the dividends in industrial productivity and prestige in the scientific community it ultimately reaped. He was also well aware that Wall Street analysts used citation counts as one measure of a firm's scientific proficiency and commercial potential. Looking back on his initial resistance to Boyer's publication mandate, Swanson remarked:

> No, it wasn't an argument or anything. It was rather, hey, we have to get the best people. How do we get them? So it all came from the philosophy, get and keep the very best people. And they were all in the academic world; how are we going to get them to come? Well, Herb said, "I know them. If we let them publish, they'll come."[80]

Yet, according to Goeddel, Swanson never completely lost his reservations about publishing research results: "Bob was always a little more worried than Herb about publications and other people knowing what we were doing. There was probably a healthy tension—Bob at one end, Herb at the other. And somewhere in the middle was how the company worked."[81]

Yet despite Boyer's insistence that Genentech would publish its work, he himself refused to appear as an author on its scholarly publications. Boyer had consulted frequently with Riggs and the molecular biologists as the insulin research progressed. To a man, they appreciated his input. But Boyer adamantly refused all entreaties to be made an author.[82]

In fact, he decided that after the somatostatin research, he would never again directly engage in Genentech's research or allow his name to appear as author on the company's scientific publications. Boyer explained the decision:

> One [reason] is I wanted to continue my own [UCSF] research, which I couldn't do at the company. . . . Another reason was I didn't want to manage a large group of scientists. I had enough of a taste of doing that at a small level to know that I didn't like it. And third, . . . I wanted to make sure that the young scientists at the company were getting the recognition. I didn't want my figurehead overshadowing anything they did. So it was a conscious decision, and I think a good one.[83]

Boyer's explanation conforms with his self-image in the 1970s as first and foremost a professor and academic scientist and also with his reputation for generosity in giving credit, especially to younger colleagues. But there is also the possibility, which Riggs acknowledged, that in light of the current barrage of criticism over his involvement with Genentech, Boyer was not anxious to further advertise his corporate ties by appearing as an author on its publications.[84]

As Swanson hoped, a spate of news coverage in the popular and scientific press played up the medical significance of Genentech's triumph, heralding human insulin as a boon to diabetics. Much of the coverage repeated the claim that the hormone, produced in efficient bacterial factories, would be purer, more plentiful, and less allergenic than the currently marketed animal insulins. Although Swanson took care to state that human insulin was far from market ready, enthusiastic media accounts tended to ignore or downplay that point.[85] *Newsweek*, for example, forecast the production of recombinant pharmaceuticals as though a sure and inevitable outcome: "The success with insulin means that recombinant DNA technology can undoubtedly be used to make scores of other vital proteins, such as growth and thyroid hormones, as well as antibodies against specific diseases."[86]

In the wake of the insulin success, Genentech's dire need was to hire staff. Its five overtaxed scientists were clearly insufficient to meet Lilly's benchmarks and also take on the new research projects the firm had planned. Boyer and Swanson, intent as ever on acquiring complementary DNA expertise, had continued to pressure Ullrich, Seeburg, and Shine to accept offers of full employment. The founders now had the insulin research as incontrovertible evidence that Genentech could successfully perform breakthrough science with results as good or better than anything academia offered. The three postdocs indeed took note of the start-up's achievements. Overcoming their initial resistance to working for a company, they began to look upon the job offers with more favor. But it was a culminating event in a string of distressing events at UCSF that disposed them to act.

In the fall of 1978, the trio learned that some months earlier the University of California had filed a patent application on the production of animal protein in bacteria, based on the biochemistry department's rat insulin and human growth hormone research. Although Ullrich, Seeburg, and Shine had contributed significantly to the work—in fact, conducted it—they found to their consternation that they were not named as inventors on the patent filing.[87] For them, it was the final straw in a series of turbulent events at UCSF in which they felt denied of scientific credit and possible patent royalties. "That [patent application] tipped the balance,"[88] Ullrich bitterly recalled. All three made a joint decision to join Genentech, affronted by their treatment at UCSF but also attracted to a firm conducting top science. Shine subsequently backed out, his wife insisting that the family return to Australia. But Seeburg and Ullrich agreed to become full-time Genentech employees, Seeburg in November 1978 and Ullrich in January 1979. Despite their commitment to Genentech, in October Ullrich and Seeburg signed yearlong consulting agreements with Lilly and continued under Lilly sponsorship to pursue their respective human insulin and human growth hormone projects at UCSF.[89] The complex entanglement of university and corporate research and alliances—and potential for conflict of interest—was clearly evident.

By the end of 1978, Genentech's total administrative and scientific staff numbered twenty-six, including the firm's first protein biochemist and several lab technicians.[90] A research organization was emerging, distinct in form and function from that of the university. Instead of academia's independent and sometimes competing university laboratories, each a hierarchical fiefdom with a professor at the top, Genentech's research beginning around this time was loosely structured into three flexible and egalitarian divisions—molecular biology, nucleic acid chemistry, and protein chemistry—all three oriented to work cooperatively to a common purpose: making marketable products. Heyneker, with customary enthusiasm, recalled the era:

> We were really, certainly in the early days, a very collegial group of people. We all had a very similar goal, of being successful, of being first, and staying ahead of the game, and not because we had to do it; because we wanted to do it. We were incredibly proud of Genentech and what was going on. We were very excited about the technology. . . . It was a terrific time.[91]

Though there clearly was teamwork and collegiality, interpersonal and interdisciplinary rivalries—inevitable in an extraordinarily competitive group of young men with their scientific reputations to prove—rumbled just below the surface. As Genentech hired organic and protein chemists, shop talk was that the molecular biologists, the lauded "cloners" and top dogs, got more kudos and credit than their fair share.

In December 1978 everyone took a brief break from the mad intensity. It had been a momentous first year as an operating company. Genentech, its laboratories completed, held an open house to inaugurate, as the invitation announced, "its newly expanded headquarters." The guest list provided a snapshot of the various constituencies that the insulin achievement had attracted to a radically new approach to making pharmaceuticals. Among the sixty or so invited were pharmaceutical industry representatives, venture capitalists, investment bankers, academic scientists, attorneys, and accountants.[92] The array of guests foretold the major participants in the new field of commercial biotechnology.

For Genentech, the consequences of making human insulin and partnering with Eli Lilly were substantial and far-reaching. The two achievements, broadly interpreted as decisive scientific and business coups, validated the company as the proprietor of a promising new technology and showcased the fact that a corporation of Lilly's stature considered recombinant DNA technology of sufficient industrial potential to warrant forming an R&D partnership with an insignificant start-up. Genentech's alliance with the pharmaceutical giant enormously magnified its visibility and boosted its chances of future financing and corporate contracts. As Middleton later observed: "That relationship [with Lilly] put Genentech on the map. That was the idea. It was an endorsement for a new technology by the leading producer of insulin in the world. It gave our company overnight credibility in the commercial world."[93] Recalling that human insulin "began the real fund-raising for Genentech," Perkins went on to comment: "We were able [after insulin] to raise money at much higher prices. So high that Kleiner and I made only token investments after that, because Kleiner & Perkins already had a significant ownership of Genentech."[94]

Important as human insulin was for Genentech's development and reputation, of wider significance was its meaning for a phenomenon about to enter popular parlance as "biotechnology." The firm's stunning synthesis of a major drug and its pathbreaking partnership with Lilly riveted the pharmaceutical sector's attention. As *Science News* asserted, achieving human insulin "catapults recombinant DNA technology into the major leagues of the drug industry."[95] With the investment window opening in the late 1970s after successive reductions in the capital gains tax, venture capitalists stepped up investment in a nascent field they hoped would prove to be as lucrative as the electronics and computer industries.[96] For example, Innoven, Monsanto's venture capital subsidiary, bought shares in Genex, a biology-based start-up founded in 1977, and International Nickel and Schering-Plough invested in Biogen. Impressed with Genentech's record in human insulin, Lubrizol Corporation, a chemical and lubricant manufacturer, made a $10 million investment in the firm, and Donald L. Murfin, head of Lubrizol's venture group, joined Genentech's modest board of directors.[97] Heeding these developments, *Nature* reported:

> Growing confidence in the US business community that the development of recombinant DNA technologies promises large profits has led to a steady flow of venture capital to support research. No one has yet made very much money, but high commercial expectations have helped raise the value (on paper) of the five small private companies most deeply involved to a figure estimated at more than $150 million.[98]

The article referenced five companies—Cetus, Genentech, Biogen, Genex, and Bethesda Research Laboratories. All except Cetus (1971) were founded between 1976 and 1978.

But it was not only the business and financial worlds that showed interest in genetic engineering. Leading molecular biology departments began to buzz with the applied opportunities foreseen in their research, the patents and licenses likely to result, and the chance for institutional and personal financial gain. It was a simultaneously exciting and troubled time of transition in molecular biology as the new genetic techniques swept university biomedical departments, supplanting previous experimental procedure and running up against academic policy and convention. Following the lead of Boyer, Cetus's Ron Cape, and Biogen's Wally Gilbert, a few entrepreneurially minded university scientists founded small biology-based companies in the general mold of Genentech, Cetus, and Biogen. David Jackson left a tenured position in biochemistry at the University of Michigan to cofound Genex in 1977. The following year, Ivor Royston and Howard Birndorf of the University of California, San Diego, established Hybritech with cofounder and venture capitalist Brook Byers.[99] Bill Rutter tried without luck to convince UCSF administrators to form a technology-transfer laboratory facilitating commercialization of campus research results. In 1981, taking matters into his entrepreneurial hands, he cofounded Chiron Corporation, a biotechnology start-up.[100] As 1978 drew to a close, the audiences for genetic engineering had considerably expanded and molecular biology was starting to lose its image of a purely academic discipline. It was entering an industrial phase in which the research direction and cultural norms of American biomedicine would be realigned to more utilitarian ends and proprietary considerations. In this reorientation, Genentech was already having a noticeable effect.

5

Human Growth Hormone: Shaping a Commercial Future

> The laboratory production of human growth hormone is . . . probably most significant for what it implies about the future possibilities in this [genetic engineering] field. If Genentech can make HGH, what else can it make?
>
> *New Scientist*, July 12, 1979[1]

After insulin, Genentech moved on without a pause to an assault on human growth hormone. The substance had figured from the start in the firm's research plans. It was on Boyer's mind as the company took shape and specified in the research and development agreements with the University of California and City of Hope. But the hormone took second place to human insulin in the firm's initial priorities—the molecule was far larger than insulin's and the current market vastly smaller. With insulin achieved, growth hormone moved to the top of the target list. Once again, Genentech scientists found themselves contending with UCSF in a fast and furious race to clone a medically relevant gene. Peter Seeburg, the UCSF postdoc destined for Genentech, was a linchpin in both growth hormone projects. He was also a disaffected participant in an explosive event—a secret midnight raid and fly-by-night escape—that would have immediate and lasting repercussions for Genentech. The incident was a high-drama outcome of the strains and stresses building at UCSF and other bastions of early recombinant DNA research as scientists glimpsed new opportunities for applied research and the fame and fortune that might attend them. Although rocked by research and personnel problems, Genentech managed to hold to a steady keel and keep steely focus on its objective—to make human growth hormone. An entirely unexpected disaster was to advance the company toward a marketed human growth hormone product and Swanson's goal of corporate independence and self-sufficiency.

COMPETING FOR HUMAN GROWTH HORMONE

Peter Seeburg had joined Boyer's lab in the spring of 1975 to begin a post-doctoral fellowship. He was one of a number of young scholars attracted to a department fast rising to molecular heights despite the precarious political environment for recombinant research. Like his fellow post-docs, he was drawn to the UCSF biochemistry department by opportunities to make a name and perhaps an academic career in a new and exciting discipline. Angular and intense, Seeburg arrived with a specific project in mind, to clone and express a synthetic gene. It was in fact the gene that Boyer expected the chemist in Germany to synthesize and send to him that fall. But disillusionment set in almost immediately. When the arrangement collapsed, Seeburg settled on growth hormone as the gene to try for, planning to use the newly breaking complementary DNA method to construct it. Finding himself at odds with Boyer, he transferred to Howard Goodman's lab, where he fell in with the young scientists jostling for recognition at the forefront of genetic engineering. It seemed a perfect environment for his growth hormone project—except for one problem. Goodman told him in no uncertain terms that the project was "too risky" and that "it might not work."[2] Quite obviously, Seeburg's supervisor was not fully supportive, a precarious situation for a postdoc.

John Baxter, a self-assured UCSF physician with dual appointments in medicine and biochemistry, had a diametrically different response. Learning of Seeburg's ambition to clone the growth hormone gene, Baxter immediately recognized a high-profile project and a scientist with the technical skills to drive it. He became the young German's enthusiastic and accommodating sponsor. But Baxter, one of the principles in Eli Lilly's contract with the University of California on insulin and growth hormone, provided more than psychological support. He supplied Seeburg with rat pituitary tumor cells, a rich source of the growth hormone messenger RNA essential for making complementary DNA copies. Seeburg spent his days doing experiments in Goodman's lab. At night, with

Baxter's permission and encouragement, he took to working secretly on growth hormone with Baxter's postdoc Joseph Martial. This aberration in academic protocol was a breach of Goodman's authority as Seeburg's supervisor, one of a series of missteps in the pressure-cooker atmosphere in UCSF biochemistry of this period. In 1977, soon after Rutter and Goodman announced their feat in cloning the gene for rat insulin, Seeburg and Martial succeeded, in another high point of early genetic engineering, in cloning (but not expressing) the gene for growth hormone in the rat.[3]

While Seeburg was celebrating publication of the research, Swanson and Boyer were deep in negotiations for funds to support a project on human growth hormone. Hans Sievertsson, the director of research at KabiVitrum—a pharmaceutical company owned by the Swedish government—had first learned of recombinant DNA technology in a discussion of Genentech's somatostatin experiment in Sweden's recombinant DNA advisory committee. Sievertsson, colleague Bertil Åberg, and their team of scientists at Kabi found the work "fantastically exciting" and wondered if recombinant DNA might be a method for producing other hormones.[4] Kabi was the world's leading commercial supplier of human growth hormone—the only form effective in humans—which it extracted from the pituitary glands of human cadavers. It was a scarce and costly drug. In high demand for treating pituitary dwarfism, the hormone was Kabi's most profitable product. But because human pituitary glands were often in short supply, the worldwide stock of growth hormone was sufficient to treat only the severest cases of pituitary dwarfism.[5] There was obvious room for market expansion—if Kabi could find a means to make the hormone in greater quantity.

Like Lilly's Irving Johnson, Sievertsson had his eye out for any new and plausible method for making the company's banner product. He found the new gene-cloning technology sufficiently intriguing to warrant flying to San Francisco late in 1977 to discuss a possible joint project on recombinant growth hormone. Since Genentech lacked a laboratory and staff at the time, Swanson must have been extraordinarily persuasive and Sievertsson extraordinarily interested, for they came to a tentative agreement to collaborate. Boyer and Swanson then immediately flew to Stockholm to continue negotiations with Kabi.[6] Their visit occurred during the heady week of the Nobel Prize ceremony in December. Roger Guillemin was one of the laureates that year for his research on peptide hormones, including somatostatin. The fact that his isolation of natural somatostatin had required grinding up tissue from almost a million sheep brains did not augur well as a basis for a commercial process. In contrast, even though Genentech had made only a minute amount of somatostatin, what impressed Åberg in particular was Boyer's vivid depiction of bacteria replicating exponentially and spewing out copious amounts of recombinant proteins. An abundant supply of human growth hormone in a market desperate for the substance meant an almost certain increase in Kabi's sales and revenues. The latter was particularly urgent for a firm reportedly teetering on the brink of bankruptcy.[7] The draw for Kabi was not recombinant DNA technology per se, but rather its potential for making a scarce hormone in greater, purer, and ultimately more profitable abundance.

That December of 1977, Genentech and Kabi signed a letter of intent, outlining the terms of a possible contractual agreement and setting out avenues for terminating the relationship if Genentech failed to meet specified research benchmarks.[8] On August 1, 1978, the two companies concluded a formal long-term research and development agreement for the use of Genentech's technology in engineering bacteria to produce human growth hormone.[9] The Genentech-Kabi contract predates the Genentech-Lilly contract by more than three weeks. It is therefore not only Genentech's first R&D agreement but the first anywhere between an established corporation and a genetic engineering firm. Yet because Kabi was not one of the pharmaceutical giants well-known in the United States, the Lilly-Genentech contract is usually taken as the original model for contractual relationships in biotechnology.[10] Furthermore, unlike Lilly, Kabi was willing to commit to a formal contract before Genentech had made the recombinant protein, perhaps an indication of Kabi's dire financial straits and desperate need to retain control of a major product. Or, more charitably, perhaps Sievertsson and Åberg had greater faith in the industrial potential of Genentech's technology.

Although the full contract has not been publicly released, as is commonplace in business practice, certain details have come to light over time. Its terms gave Genentech twenty-four to thirty months from the August 1978 signing to develop bacteria producing human growth hormone. Åberg later reported (and as a signatory to the agreement was in a position to know) that Kabi agreed to pay Genentech $1 million for engineered bacteria producing human growth hormone.[11] The contract also gave Kabi exclusive foreign marketing rights, but with Kabi and Genentech sharing rights in the United States. Kabi was to provide Genentech with human pituitary source material and send protein chemists and fermentation experts to South San Francisco to collaborate in creating a production process. Swanson insisted that the contract made clear that Kabi was to apply Genentech's engineered bacteria solely for the purpose of making growth hormone. As in the Lilly contract, he was willing to sell know-how and biological material for specific applications. But he adamantly refused, as a hard-and-fast rule, to sell Genentech technology for other than the uses the contracts spelled out.

With Sweden's controversy over recombinant DNA research rising to a crescendo, Kabi's management at first kept secret its collaboration with Genentech and adoption of genetic engineering. Then in 1978, it spun off the technology into a new and separate company called KabiGen, in part to avoid the tension inside and outside the parent company regarding application of a suspect technology.[12] Lilly's contracts with Genentech and the University of California, Kabi's contract with Genentech, and Biogen's contracts with Harvard and other universities show commercial interests overriding contrary politics and adverse public opinion. It was a trend that would soon gain momentum.

In the summer of 1978, Seeburg and his UCSF colleagues achieved another milestone, actual expression of growth hormone in the rat.[13] Boyer and Swanson immediately stepped up their effort to recruit him, plying him with sailboat rides and barbecue picnics to entice him. Seeburg's disgruntlement with circumstances at UCSF and the opportunities he saw at Genentech finally convinced him to join the company. With Genentech's human insulin coup that August, his confidence grew that the company provided the proper environment for applied research in biology. Swanson put it concisely: "I think both [Seeburg and Ullrich] felt that Genentech was the best atmosphere to actually get the thing done, that they could move more quickly at that point in a corporate environment than an academic one."[14] In September Seeburg signed Swanson's recruitment letter and gave the university notice of his departure at the end of October. Genentech had acquired in the persons of Seeburg and Ullrich the much sought-after expertise in complementary DNA cloning.

Wasting no time, Genentech's still-skeletal scientific team and its City of Hope colleagues had moved instantly from human insulin to the new project on growth hormone. On September 12, less than a week

Fig. 15. Dennis Kleid, Dave Goeddel, Art Levinson, Herb Heyneker, Peter Seeburg, Dick Lawn, and Axel Ullrich, Pajaro Dunes, California, September 1982. (Photographer unknown; photograph courtesy of Herbert L. Heyneker.)

after the insulin news conference, both scientific groups plus Swanson met at Genentech to devise a research plan. Seeburg and John Shine, still UC employees, also attended. (Goeddel was rock climbing—his manner of relaxing after the ardors of the insulin contest.) Because the growth hormone molecule is almost four times larger than insulin's—191 amino acids compared to insulin's 51—the scientists concluded that chemical synthesis of the DNA coding for that many amino acids would require an inordinate amount of time and labor. Crea later estimated that to synthesize the complete gene using the technology of that era would have taken one to one and a half years—far too long and costly for a company premised on speed of execution and frugality.[15] The methodology of the human insulin experiment was clearly impractical for making growth hormone.

The scientists then proposed a highly original concept: they would attempt to build a semi-synthetic gene coding for growth hormone, using DNA synthesis and complementary DNA methods. Asked in litigation two decades later who came up with the strategy, Seeburg replied, "Oh . . . this is difficult to say. I don't think I can name one particular person that had this idea."[16] The patent application of July 5, 1979, nonetheless named

Goeddel and Heyneker as inventors. The City of Hope chemists were to construct a short segment for the front of the artificial gene, including the bacterial signal system controlling gene expression. Seeburg was to make a long complementary DNA segment coding for most of the human growth hormone protein. The plan was to clone the synthetic and complementary DNA pieces in separate procedures and then join the two to form a hybrid gene. Making a functional semi-synthetic gene was a tall order; it had never been done before and would require a new level of technical sophistication. Two days after the meeting, Swanson sent Seeburg, and presumably also John Shine, checks for their consulting services.[17] Seeburg and Shine, along with Howard Goodman, were also consulting on growth hormone for Eli Lilly, just as Rutter, Goodman, and Ullrich were consultants for Lilly on insulin. The UCSF biochemists had clearly grasped the applied potential of their research and appeared willing to take on conflicting alliances.

A strategy in place, Itakura's lab immediately began the tedious work of synthesizing the DNA coding for the first 24 amino acids. That accomplished, the research devolved to the Genentech molecular biologists. Goeddel and Heyneker linked the synthetic DNA fragments for the front of the gene and began to construct a plasmid to express the hybrid gene. Everyone expected Seeburg, with his arrival at Genentech in November, to start immediately constructing the complementary DNA component. Instead, the project rapidly unraveled. What happened next speaks to the raw intensity and extraordinary competitiveness of recombinant DNA research of the late 1970s.

A letter may have inadvertently set the ball rolling. In November 1978 Swanson and Kleid wrote to Goodman, requesting the transfer of Seeburg's biological materials to Genentech. They specifically requested the precious human growth hormone complementary DNA that Seeburg had made at UCSF.[18] Goodman flatly refused. Three months later, tension rising, Goodman and John Baxter warned Seeburg that any research materials removed without the university's authorization and used in experiments violated UC policy and the NIH guidelines.[19] The impetus for the letter was very likely an extraordinary event, extraordinary even in a decade of extraordinary events in the formerly staid field of molecular biology.

The incident or "midnight raid," as Ullrich referred to it, occurred on New Year's Eve 1978 as he made final preparations to go to Genentech.

United States Patent [19]

Goeddel et al.

[11] **4,342,832**

[45] **Aug. 3, 1982**

[54] **METHOD OF CONSTRUCTING A REPLICABLE CLONING VEHICLE HAVING QUASI-SYNTHETIC GENES**

[75] Inventors: David V. Goeddel; Herbert L. Heyneker, both of Burlingame, Calif.

[73] Assignee: Genentech, Inc., South San Francisco, Calif.

[21] Appl. No.: 55,126

[22] Filed: Jul. 5, 1979

[51] Int. Cl.³ C12N 15/00
[52] U.S. Cl. 435/172; 435/68;
435/70; 435/317; 536/27
[58] Field of Search 435/317, 172, 68, 70,
435/71

[56] **References Cited**

FOREIGN PATENT DOCUMENTS

6694 6/1979 European Pat. Off. .
20147 12/1980 European Pat. Off. .

OTHER PUBLICATIONS

Technology Review, pp. 12 and 13, Dec. 1976.
Martial et al., Science, vol. 205, Aug. 10, 1979.
Shine et al., Nature, vol. 285, Jun. 12, 1980, pp. 456–461.
The Economist, pp. 87 and 88, Jul. 14, 1979.
Time, Jul. 30, 1970, p. 70.
Newmark, Nature, vol. 280, pp. 637 and 638, Aug. 23, 1979.
Villa-Komaroff et al., Proc. Natl. Acad. Sci., vol. 75, pp. 3727–3731, Aug. 1978.
Seeburg et al., Nature, vol. 276, pp. 795–798, Dec. 1978.
Itakura et al., Science, vol. 198, pp. 1056–1063, Dec. 1977.
Crea et al., Proc. Natl. Acad. Sci., vol. 75, pp. 5765–5769, Dec. 1978.

Klenow et al., Proc. Natl. Acad. Sci., vol. 65, pp. 168–175, Jan. 1970.
Sutcliffe, Cold Spring Harbor Symposium 43, pp. 70–90 (1978).
Curtis et al., Molecular Cloning of Recombinant DNA, by Scott et al., pp. 99–111 (1977).
Ullrich et al., Science, vol. 196, pp. 1313–1319, Jun. 1977.
Bolivar et al., Gene 2, pp. 95–113 (1977).
Goeddel et al., Proc. Natl. Acad. Sci., vol. 76, pp. 106–110, Jan. 1979.
Chang et al., Nature, vol. 275, pp. 617–624, Oct. 1978.
Maxam et al., Proc. Nat. Acad. Sci., vol. 74, pp. 560–564 (Feb. 1977).
Kornberg, DNA Synthesis, pp. 87 and 88, pub. by W. H. Freeman & Co., 1974.
Razin et al., Proc. Natl. Acad., vol. 75, pp. 4268–4270, Sep. 1978.
Wickens et al., The Journal of Biological Chemistry, vol. 253, No. 7, pp. 2483–2495 (1978).

Primary Examiner—Alvin E. Tanenholtz
Attorney, Agent, or Firm—Thomas D. Kiley

[57] **ABSTRACT**

Described are methods and means for the construction and microbial expression of quasi-synthetic genes arising from the combination of organic synthesis and enzymatic reverse transcription from messenger RNA sequences incomplete from the standpoint of the desired protein product. Preferred products of expression lack bio-inactivating leader sequences common in eukaryotic expression products but problematic with regard to microbial cleavage to yield bioactive material. Illustrative is a preferred embodiment in which a gene coding for human growth hormone (useful in, e.g., treatment of hypopituitary dwarfism) is constructed and expressed.

12 Claims, 5 Drawing Figures

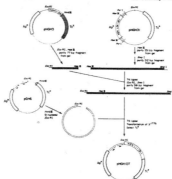

Fig. 16. First page of the 1982 patent on Genentech's semi-synthetic growth hormone method, 1982. United States Patent and Trademark patent databases.

Seeburg, whom Goodman had banned from the premises after a furious dispute in November over his ties to Genentech, asked to accompany Ullrich to remove some biological samples and take them to the company across the bay.[20] Ullrich agreed, being "not very sympathetic" to Goodman at the time because of what he and Seeburg felt were instances of misappropriated credit, particularly the failure to include them as inventors on the patent application.[21] Around midnight, the two entered the deserted lab and removed various research specimens, including some of Baxter's human pituitary material and a complementary DNA clone of human growth hormone. They drove straight to Genentech, where, flashing their employee badges, they breezed past a security guard, entered the building, and deposited the materials in a freezer.[22] The escapade was a stunning expression of the stresses rife in recombinant research of the period as participants battled for scientific prestige, industrial contracts, and future patent royalties.

It was not uncommon at the time for scientists to take their research materials with them to a new institution, often without higher authorization. Nonetheless, it was a gray area of academic practice with few clear guidelines. Adding further complexity was the fact that Ullrich and Seeburg had taken the specimens not to another university, but to a private, for-profit company. They soon found themselves trying to justify their actions to astonished colleagues and the press. In February 1980 a defensive Seeburg told a journalist:

> I had largely started the growth hormone project and worked on it since 1975. Why shouldn't I take material which I had acquired? I didn't take anything exclusively. Whatever I took I left some behind. So we all had the same starting point after that. It wasn't that I had any edge except I knew in my head what experiments I would do next. I feel I just went to another lab and continued my work.[23]

Goodman, Baxter, and the patent administrator in the Board of Patents Office at the University of California demanded the return of the materials in a chain of increasingly threatening letters.

At issue was far more than an investigator's right to take research materials to another institution, although that indeed was an issue. Much more important was the question of proprietary rights to the transferred material, a particularly significant issue in light of the university's patent application and Genentech's plan to file for patents on its growth hormone research.[24] In April 1979 the university's patent administrator wrote to Swanson, stating in no uncertain terms that the materials Seeburg and Ullrich had removed to Genentech were the subject of patent applications and belonged to them "in their personal capacity only, and not to any commercial entity. The fact that Drs. Seeburg and Ullrich are currently employed by Genentech, Inc. shall not be construed as giving Genentech any interest in said materials whatsoever, absent express permission from The [UC] Regents."[25] It was one of several warnings by the university that any uses Genentech made of the biological materials Seeburg and Ullrich had brought to the firm were unauthorized. The issue would simmer ominously for almost two years, coming to a temporary settlement in 1980 in Genentech's run-up to a public stock offering, as chapter 6 relates.

While Genentech and the university wrangled over the transferred materials, Genentech's scientists had problems of their own. The growth hormone project had ground to a standstill. Seeburg made little if any progress after several months of on-again/off-again attempts to clone the complementary DNA sequence. The novelty of the science was not the only problem; he was overcome by drug, alcohol, and marital problems, as he later admitted in court.[26] He showed up at work for a few hours at best and made little progress. Swanson, fretting over the stalemate, imagined the Baxter-Goodman team pulling ahead and winning what had become a bona fide race for human growth hormone. All the while, rumors circulated ominously in the biomedical community. As one publication put it: "[Seeburg's] recent move from UCSF to Genentech has been accompanied by questions in the scientific community whether he took with him more tangible substances than the knowledge of recombinant techniques that he acquired while working for a publicly funded institution."[27] It was a tricky situation, as Yansura recalled: "[Swanson] didn't want to irritate Seeburg or push him out because Peter Seeburg had [the skill to make] the growth hormone [cDNA] gene, and wherever Peter went the growth hormone gene went. That would be a bonus to our competitor."[28]

To rescue the stalled project, Swanson turned to Goeddel, fast becoming Genentech's prized cloner. In February 1979 Goeddel, with Heyneker assisting, set out to construct and clone the complementary DNA segment. It was tough going. One problem was the unsatisfactory condition of the human pituitary material that Kabi provided as a source of RNA for making the complementary DNA segment. "Goeddel struggled because he tried to make high-quality cDNA with poor-quality messenger RNA,"[29] Heyneker observed. In July Goeddel and Seeburg succeeded in cloning the complementary DNA segment, finally bringing to a halt months of frustrating work. Whether they used the growth hormone complementary DNA Seeburg had made at UCSF would become a point of bitter litigation between UC and Genentech in the 1990s.[30] Goeddel then enzymatically linked the synthetic DNA and complementary DNA pieces, inserted the hybrid genes into expression plasmids, and transferred them to bacteria for cloning and expression.

The first sign of success was Goeddel's whoop as he checked the scintillation counter: "It's pinned! It's pinned!"—shop talk meaning the reading for the radioactively tagged growth hormone protein had soared completely off the chart.[31] It was strikingly clear, as Heyneker recalled, that the bacteria were making "a significant amount of growth hormone right off the bat."[32] Pandemonium erupted when they showed the results to Swanson. People streamed in from all directions, exchanging high fives with the triumphant team. Genentech had made a complicated gene and induced bacteria to churn out pure human growth hormone in quantity—200,000 molecules per bacterium, under optimal conditions, according to Goeddel.[33] Kabi had bet on a dark horse and won.

To Genentech scientists, worried that somatostatin and human insulin might be mere flashes in the pan, achieving growth hormone meant more than the successful synthesis of another recombinant substance. The company's first two projects had fallen short of establishing that its technology was widely applicable for the bacterial production of useful proteins. Everyone recognized the fusion-protein approach as cumbersome and restricted to making a narrow range of substances.[34] A nagging concern was that Genentech could not build a sustained business on the limited utility of the fusion-protein approach. That concern evaporated with the growth hormone triumph. Genentech's hormone was not a fusion protein; it was not a precursor; it was a pure and freestanding substance, and it was produced in relative abundance. Lab tests showed it to be biologically active and identical to the natural hormone. It all added up to the creation of a seemingly versatile and efficient model for making the larger and more complex proteins in common clinical use. Substances such as the blood-clotting factors with genes too large to synthesize now seemed within eventual grasp. One perceptive reporter remarked, "The laboratory production of HGH is . . . probably most significant for what it implies about the future possibilities in this [genetic engineering] field. If Genentech can make HGH, what else can it make?"[35] In short, the making of growth hormone indicated a far clearer path, albeit with inevitable detours and impediments, to a viable commercial future.

The elation sweeping Genentech that day might have fallen a notch if the group had known that UCSF's growth hormone team was close on its heels. That summer Baxter, with Eli Lilly's money, had sent Joseph Martial to the same Lilly-owned laboratory in France in which Ullrich had cloned the human proinsulin gene. Once again, the aim was to escape the more stringent U.S. restrictions on recombinant DNA research. The NIH guidelines, designed to reduce perceived hazards, served paradoxically in these and a few other cases to move experiments and their hypothetical dangers to other countries. By June 1979 Martial had succeeded in cloning and expressing a complementary DNA segment for human growth hormone.[36] Although a notable achievement, he had produced growth hormone as a fusion protein that remained biologically inactive until stripped of the attached bacterial protein.

Swanson engaged a public relations firm to prepare a press notice. The result was the usual promotional text. But in this instance it came complete with a sketch of the research procedure and a diagram of the growth hormone molecule coiling sinuously around the page, its lengthy string of amino acids sequentially laid out as labeled beads.[37] Genentech's notice was scheduled for public release on July 12, 1979, after a patent application and a paper for journal publication were in the mail and Goeddel had reported the work at a symposium. Then suddenly the timing of the announcement disintegrated. Alerted by reporters to Genentech's imminent announcement, John Baxter on July 10 made a rushed verbal report on the UCSF team's production of human growth hormone. Genentech, not to be out-competed, countered the same day with its own oral report. Swanson then moved the release of the written press announcement up a day, to July 11.[38] It was one-upmanship worthy of playground rivals but with far higher stakes.

With the élan typical of corporate announcements, and increasingly also of university publicity, Genentech's news release declared the production of growth hormone in unattached form "a major milestone."[39] The author of the release took pains to distinguish the product from the fusion proteins made in the somatostatin, insulin, and UCSF growth hormone work. Genentech's aggressive new marketing manager delighted in telling a reporter: "What we're producing is human growth hormone. What John Baxter's lab is producing is growth hormone attached to something else. And the clinical efficacy of that substance is completely unknown."[40] Baxter countered, his competitive streak in plain evidence. He observed to the press that supplies of natural growth hormone had never been sufficient to treat children failing to grow normally. Putting his native southern charm aside, he took a stab at Genentech, remarking, "This [growth hormone] is a hormone we really need, unlike insulin"—which he considered in adequate existing supply.[41] In fairness, the objectives in the competing projects were slightly different. Genentech wanted first and foremost a commercial growth hormone product; the UCSF group, in no way adverse to producing a marketable hormone, was nonetheless primarily concerned to elucidate the mechanisms behind mammalian gene expression in bacteria. However interpreted, Genentech in its triple gene clonings—somatostatin, human insulin, and growth hormone—had demonstrated that top-flight biology was no longer the sole province of academe. Clearly, Genentech had moved into the hallowed circle, with other upstart companies to follow.

Reporters focused on the growth hormone research as a pharmaceutical breakthrough, seeking to appeal to a readership interested in medical advances. Articles made much of bacterial "factories" overcoming former hormone shortages, but devoted less ink to noting a manufacturing process yet to be worked out and a product far from the marketplace. Swanson, however, was quoted as cautioning the public that Genentech and Kabi had yet to develop a production process, let alone come close to gaining FDA product approval.[42] The *Economist*, in an article callously titled "No More Dwarfs," commented on the speed with which Genentech had created "three new medical products in as many years."[43] The company had indeed succeeded in cloning three genes faster than anyone expected. But it was a leap of faith to refer to the recombinant proteins as "medical products" when the long industrial development stage and uncertainty of FDA approval yawned before them. The media that a few years earlier had tended to dwell on the possible hazards of genetic engineering were increasingly extolling its scientific achievements and practical contributions.[44] More significantly, governmental bodies created to monitor DNA research were beginning to lament regulatory roadblocks to full basic-science and industrial exploitation of the field.

Swanson soon came to see growth hormone as more than a cloning success and a likely future product. He believed that only by making and
selling its own pharmaceuticals could Genentech capture full monetary
value from the heavy cost of pharmaceutical research and development.
As he observed two decades later:

> Over the long run—and really the timing is when you can achieve it—in
> order to capture all the value from the research that develops a new drug
> that treats a disease, you have to be able to make and sell that drug your-
> self, in part to control the distribution of it, not relying on someone else;
> and in part because you capture greater rewards by selling it yourself. Over
> the long run, unless you capture those rewards, you cannot invest as much
> in R&D that allows you to develop the second and third products.[45]

Taking on drug development, approval, and marketing was a monumental challenge for a start-up with no products generating income, no
deep investor pockets, and just over fifty employees.[46] A more realistic approach, Swanson decided, was to tackle corporate integration stepwise,
research project by research project. With an incremental approach, Genentech might build up its own manufacturing capacity and break away
from reliance on a pharmaceutical company for product development
and approval. He reasoned that creating a comprehensive internal research and development program would compel Genentech to obtain the
knowledge and resources to practice the full range of expertise entailed
in bringing a pharmaceutical to market. Clearly, Swanson had no intention of the firm remaining solely a contract research operation; he had
more expansive ambitions and little patience for tarrying.

For several reasons, growth hormone seemed an appropriate project
for Swanson's scheme. The hormone had no entrenched competition in
the United States: the nonprofit, government-supported National Pituitary Agency, created in the 1960s to collect pituitary glands from coro-
ners and private donors, dispensed growth hormone free of charge to
physicians treating children with severe forms of pituitary dwarfism.
Swanson saw the circumscribed distribution of growth hormone as an
advantage. Genentech should be able to get by with a small sales force
since only a few physicians (largely pediatric endocrinologists) currently

prescribed the hormone. By selling directly to them, the firm could avoid the immense cost of drug promotion and distribution.[47] The agency believed supplies to be sufficient for current treatment but welcomed the possibility of additional low-cost growth hormone so that exploratory research on new applications could proceed.[48] Perhaps there was room for modest market expansion after all. Yet no one at the company, including Swanson, anticipated a large market for growth hormone. An inside joke was that sales revenues might cover the cost of the firm's toilet paper.

His mind made up, Swanson approached Kabi about amending the contract. In 1980 he succeeded in licensing from Kabi exclusive rights for Genentech to sell recombinant growth hormone in the United States. In return, Genentech reduced the royalty rate Kabi was to pay Genentech on foreign sales.[49] With the American market now Genentech's exclusive purview, the pressure was once again on the scientists. Crea's lab, up and running at Genentech by the end of 1978, filled unending requests from the other groups for synthetic DNA sequences and probes; the molecular biologists labored to increase growth hormone yields; and the first protein chemists to arrive at the firm strove to improve hormone purity to the high standard required for clinical trials. As 1979 drew to a close, Swanson claimed, with more than a little exaggeration, that Genentech was already "a fully integrated company engaged in the research, development, manufacture and marketing of commercially valuable substances produced by specially engineered microorganisms."[50]

By then, Swanson had come to a different opinion about the potential size of growth hormone sales. As early as the fall of 1979, he began to predict an expansive market with a variety of potential clinical uses for the hormone. What had started as a project aimed primarily at advancing the company's technological capacity and corporate evolution, he now came to regard as having substantial commercial potential. Anticipating clinical applications beyond the treatment of pituitary dwarfism, he began to publicize the financial windfall likely to result from growth hormone sales. In September 1979 he told a group of stock market analysts at the brokerage firm E. F. Hutton that achieving a genetically engineered form of human growth hormone in the unprecedented quantities expected would allow investigation of its uses in treating wounds, bone fractures, and other medical conditions. Predicting a market of over $100 million if these additional indications materialized, he told the analysts that "HGH, as it is called, may turn out to be one of the most important substances yet to be produced via genetic engineering."[51] Swanson, it was obvious, no longer regarded growth hormone as a small-market product; in his opinion, it now had signs of huge sales potential. For that actually to materialize, Genentech needed to transform a small-scale lab procedure into a productive and efficient manufacturing process.

SCALING UP INSULIN AND GROWTH HORMONE

Genentech's modest scientific staff was already contending with the development of human insulin to meet Lilly's stringent requirements. Achieving the hormone had been remarkable, but the experiment had produced only an infinitesimal quantity. In comparison, the growth hormone procedure expressed the hormone in relatively substantial amounts, but even so it fell far short of constituting anything close to an industrial production process. An open question was whether Genentech could develop laboratory-scale procedures into a platform for manufacturing proteins in the amounts and purity meeting FDA and market requirements. Yansura recalled the worry:

> There was a period where we were wondering whether it was really possible [to develop recombinant products commercially], and if it wasn't possible then we were all in this dream job that wasn't going to last. Our job in the industry was based on the fact that we could make proteins in a large enough scale to be able to sell them.[52]

Swanson had begun in 1979 to hire a cadre of process engineers and fermentation experts to handle the development and scale-up of insulin and growth hormone. It was they and technologists at Lilly and Kabi who were responsible for developing the first pharmaceutical manufacturing systems for recombinant pharmaceuticals.[53] Integrating laboratory-scale approaches into viable development and production processes was anything but easy, requiring innovation and radical adaptation of existing procedures. As Boyer recalled:

> The absolutely amazing thing to me was the manufacturing component of the industry which had to be developed. It was totally new. It wasn't fermenting beer. It wasn't making antibiotics. It was completely different. It was engineering organisms to make a unique protein, which, in turn, could

be purified and expressed in large quantities by the organism that was engineered, and to provide this in large enough quantities to do the clinical studies and eventually to make it available to physicians for clinical use.[54]

But trouble lay ahead. It was in the course of increasing the volume of bacterial cultures in the scale-up process that Genentech collided head-on with the NIH guidelines, which up till then it had largely avoided.

Swanson from the start had found it prudent and politic for the firm to comply with the guidelines even though they applied only to institutions receiving NIH research funding. In light of the political controversy, he was eager for the public to see the firm as a responsible corporate citizen totally in compliance with the guidelines. Swanson had asked Dennis Kleid to establish a biosafety committee modeled after the one he had founded at the Stanford Research Institute. Kleid then proceeded to form a committee of Genentech personnel to which scientists with projects involving natural DNA, including complementary DNA, submitted proposals for review. Swanson signed off on approvals. As NIH policy evolved, the committee added a mandatory external member.[55] At first, the guidelines had minimal impact on the company's research; its experiments involved primarily synthetic DNA that the original guidelines did not cover.

As Genentech's insulin and growth hormone projects began to expand beyond bench-top experiments, Swanson's ideas regarding the guidelines shifted. At the time of the somatostatin announcement in December 1977, he had stated publicly that Genentech's adherence to the guidelines was a top corporate priority. As the first company to industrialize recombinant DNA, Swanson thought "it important to set a good example" in strict voluntary compliance with the guidelines.[56] A year later Swanson had changed his mind. Faced with providing Lilly with bacterial cultures producing insulin in specified quantity and purity, his priority had become meeting the benchmarks—even if it meant flaunting aspects of the guidelines. He and Boyer had even gone to Washington, DC, to lobby against legislation regulating recombinant DNA research, arguing that the industrial prospects of the research should not be reined in.[57] With the pharmaceutical industry routinely employing microbial cultures many times the guideline limit, the ten-liter maximum was an obvious roadblock to industrial manufacturing in and commercialization of a promising new field.

In June 1979 Swanson brought his objection to the ten-liter limit out in the open—perhaps hoping to play on the public's eagerness for new and more effective drugs. He pointedly informed the media that Genentech's current volume of insulin cultures was around sixty liters—six times the NIH guideline limit.[58] The *Washington Post* noted that the escalation in batch size without formal NIH approval "emphasizes industry's impatience with the guidelines."[59] Swanson was indeed impatient. Shrewdly stressing medical rather than business priorities, he told the *Post* that Genentech had an urgent obligation to provide human insulin to the diabetic population.

> Because we've been leaders in this technology, we've had to deal with the problem of how to get the insulin out to the diabetics in this country. We're now working as fast as we can to produce enough for animal tests in the fall. You have to do things on a larger scale to make the test.[60]

Swanson's remarks had an obvious political subtext: the health and welfare of the diabetic population trumped any other consideration, including what he saw as ill-conceived federal policy restricting the manufacture of recombinant products. In regard to growth hormone, Genentech and Kabi employed a less confrontational strategy to circumvent the ten-liter limit—and the Swedish government's similar mandate. Kabi's and Genentech's process engineers adjusted work schedules to run fermenters in continuous cycles at a maximum of ten liters.[61]

All the while, Lilly's Irving Johnson, a level but forceful voice in opposition to stringent federal controls, petitioned the NIH through the Recombinant DNA Advisory Committee to remove the ten-liter batch-size limitation and make other procedural changes favoring industry. Government policy, he argued, should not cripple through restrictive policies and time-consuming procedures the industrialization of a technology with impressive evidence of medical and economic significance.[62] By 1980 U.S. regulatory clamps were loosening. Washington had turned to technology-based companies as one avenue for lifting the nation out of recession and placing it on a competitive footing on the international stage. As an element in a general reduction of industrial regulation under the Reagan administration, the NIH that year relaxed guideline strictures—lowering or removing containment requirements for many types of experiments. The NIH now permitted large-scale experiments (above ten liters) on a case-by-case basis if, as a government report re-iterated, the recombinant DNA was "rigorously characterized and the absence of harmful substances established."[63] The report, published in 1981, went on to claim that 85 percent of recombinant DNA experiments could now be performed at the lowest containment levels—essentially ordinary lab conditions.

In January 1980 NIH director Donald Fredrickson took special action to approve the seven requests Genentech and Lilly had submitted for conducting cultures of engineered organisms above ten liters. Human insulin and growth hormone were conspicuous among the approvals.[64] Not everyone was pleased. Some expressed alarm over the prospect of Lilly and other corporations mass-producing vats of recombinant organisms in the absence of meaningful government oversight and regulatory enforcement.[65] But industry could claim a victory of sorts: the Recombinant DNA Advisory Committee agreed in mid-1980 to amendments that made large-scale fermentations permissible without the NIH director's prior approval.[66]

With the way now clear to produce substantial quantities of human insulin, in 1980 Lilly began the first of a series of clinical trials in human volunteers and also launched construction of two immense factories for manufacturing the recombinant hormone.[67] Swanson had predicted in the 1976 business plan that insulin would not undergo long regulatory delays because, as he put it, "insulin is not a new drug." He predicted that it should consequently sail through the clinical testing and FDA approval processes.[68] He was largely right. The regulatory approval process for insulin went forward without a major hitch. But Lilly's long experience with shepherding the animal insulins and other biologicals through the FDA, plus its long-term ties with drug regulators, were also definite advantages. In October 1982 the FDA approved the sale of the Genentech-Lilly insulin, under the trade name Humulin. It was the first recombinant pharmaceutical for human use to reach the marketplace.[69] The trajectory from lab bench to product had taken a mere four years—less than half the time onlookers had originally predicted for commercializing recombinant DNA technology. As Irving Johnson took pleasure in remarking, "We must all be impressed with the speed with which the technology has progressed since 1974."[70]

The development of human growth hormone into a commercial product took a longer and more erratic course. In January 1981 Genentech announced FDA approval to proceed with human clinical trials.[71] What the announcement didn't mention was that a handful of Genentech employees would first receive injections of growth hormone to test for safety. (A safety study in employees would violate today's far-stricter protocol for clinical trials in humans.) Recipients suffered soreness and fever, found subsequently to be due to bacterial contaminants in the formulation, so the clinical trials were postponed.[72] The setback introduced another period of intense labor to achieve a purer product. That attained, Genentech reported in September 1981 that ten medical institutions were set to run clinical trials.[73] By winter 1982 Genentech, confident that it had a growth hormone product in the offing, was close to completing a 72,000-square-foot manufacturing facility in its South San Francisco complex.[74] But the road ahead proved rocky indeed. The FDA balked at the information that Genentech's recombinant hormone was not identical to the natural molecule and had elicited adverse antibody responses in clinical trials.[75]

Then the entirely unexpected occurred. In spring 1985 four adults who as youngsters had received injections of cadaver-derived growth hormone died of Creutzfeldt-Jakob disease, an incurable neurological syndrome akin to mad cow disease.[76] The FDA and several European regulators reacted to the calamity by requiring distributors, including Kabi, to withdraw the natural hormone, suspected to be contaminated with an unknown pathogenic agent.[77] Genentech swiftly issued a press release to assure the public that its recombinant growth hormone was not derived from human tissue and was pure and safe.[78] In light of the medical emergency, the FDA precipitously dropped its reservations. On October 18, 1985, the FDA approved Genentech's human growth hormone for sale in the United States under the trade name Protropin. Genentech almost entirely on its own had engineered, developed, and won regulatory approval of what would become an immensely lucrative product.[79] But to an elated Swanson, the company had done more than that: "With today's approval, Genentech achieves a major milestone we established at our founding—to market our own products."[80] To his mind, the company had met a major goal set at its foundation.

Even *Business Week* acknowledged that Genentech had taken "the first step to becoming a full-fledged pharmaceutical company." The reporter, obviously impressed, went on to describe the company's exuberant response to the FDA approval: "When things go right at Genentech Inc., the scientists have a novel way of expressing their excitement. They face off, jump in the air, exchange overhead hand-slaps, and cry, 'DNA!' . . . The company's co-founder and chief executive, Robert A. Swanson, happily admits that he's been throwing a few high-fives himself lately."[81]

Eight days after the public announcement, Genentech pulled out all stops in a lavish celebration. The entire company assembled to dine and dance under an enormous tent the size of a football field set up in one of its parking lots. So brilliant were the accompanying fireworks that officials at neighboring San Francisco International Airport momentarily suspended air traffic. Swanson announced that all employees, except corporate officers, would receive options to purchase one hundred shares of Genentech stock.[82] *Business Week* later pictured Swanson on its cover in a white lab coat, with a satisfied grin on his face, against the background of a manufacturing facility. Block letters above the photo blasted, "Biotech Superstar," and slightly more modestly underneath, "Wall Street Loves Genentech."[83]

The company had been proactive in taking steps to move quickly into the market once Protropin was approved: it had stockpiled the hormone for close to a year and, having mined the pharmaceutical industry for seasoned salespeople, had a marketing team assembled and ready to go.[84] Within a week of FDA approval, Genentech began to ship growth hormone to hospital pharmacies around the country.[85] In another unanticipated event, in December 1985 Protropin received orphan drug status under the Orphan Drug Act of 1983. The act gave seven-year marketing exclusivity and generous tax credits to companies making drugs for rare disorders with limited market potential.[86] As the cadaver-derived hormone slipped from favor, revenues from recombinant human growth hormone would turn out to be far more magnificent than Swanson originally projected. Within two decades, Protropin sales reached $2 billion.[87]

CORPORATE EXPANSION

As the 1970s drew to a close, Genentech was undergoing expansion in scale and complexity. By the second quarter of 1979, the company had four new projects under way, all but one sponsored by a major corporation: Hoffmann–La Roche on interferon; Monsanto on animal growth hormone; Institut Mérieux on hepatitis B vaccine; and a Genentech-funded project on the hormone thymosin.[88] The demands of these projects, on top of Genentech's obligations to Lilly on insulin and the in-house manufacture of growth hormone, called for enlargement at the management level. In January 1979 Robert Byrnes arrived to become Genentech's first vice president of sales and marketing. Byrnes had decided to leave a well-paid position as an American Hospital Supply marketing executive for the singular opportunity to build an entirely new marketing division at Genentech. Tom Kiley, succumbing to infatuation with Genentech technology and the legal issues it raised, joined the firm full-time as general legal counsel in February 1980. He was the only individual to become employed at the early company ostensibly risk-free: his former law partners, apparently fearing for his future at the renegade firm, guaranteed to reinstate him at Lyon & Lyon if his position at Genentech failed to work out.[89] (The precaution proved unnecessary; Kiley of his own volition retired from Genentech in 1988, in several ways far richer for the experience.) Later in 1980 Bill Young left the security of a longtime position in process engineering at Eli Lilly to become director of manufacturing at Genentech.[90] Despite Lilly's every attempt to keep him, he found the chance to be at the forefront of creating manufacturing procedure for recombinant pharmaceuticals irresistible. With Young's arrival, Genentech's first full management team was in place.

The new hires soon learned that job titles fell short of describing the full scope of their responsibilities. They found themselves performing any and all tasks required to keep research and the company moving. Byrnes described what he learned to expect, a far cry from the circumscribed job descriptions of the pharmaceutical industry: "You have to be flexible . . . and not overly concerned about what you do day-to-day— whether it's running out to get a liver for a scientist or playing the role of vice president in a negotiation. It doesn't matter. The point is, I'm a resource."[91] Rigid business organization and sharply delineated functions had no place at Genentech, a company in which flexibility, improvisation, and quick action were essential. Years later Bob Byrnes recalled that Genentech's business charts, if they existed at all, failed to reflect the company's actual organization. That disparity became a problem in 1990 when Hoffmann–La Roche, engaged in a 60 percent acquisition of Genentech, demanded a formal organization chart. Byrnes had to hustle to create a chart more accurately depicting corporate reality.[92]

Young as Swanson was—younger than his management team and most of the scientists—he was firmly at the helm and holding Genentech to a tight rein. He kept everyone focused on product-oriented research, meeting benchmarks, and creating corporate value through alliances, investment, products, and patents. His initial objection to the somatostatin project was not an aberration; he continued to have scant tolerance for spending time, effort, and money on research not tied directly to producing marketable goods. Focus on products had become a mantra that everyone could repeat in their sleep. Dan Yansura put it in a nutshell: "We were interested in making something usable that you could turn into a drug, inject in humans, take to clinical trials."[93]

A few years before his premature death in 1999, Swanson remarked, "I think one of the things I did best in those days was to keep us very focused on making a product."[94] His goal-directed management style differed markedly from that of Genentech's close competitors. Cetus, a company of multiple visions, seemed unsure of what exactly it wanted to become. Its president Ron Cape admitted as much: "We were looking for credibility, and we didn't have a fixed business model. We shifted from [projects with] Schering[-Plough] to Chevron and Amoco because that's what showed up. . . . There was a certain *ad hoc* aspect to what we did."[95] Biogen, with its research parceled out among several prestigious European and American laboratories—each headed by a strong personality—had no centralized decision-making structure, no regular and productive interactions, little concentrated executive authority. Biogen's lack of a common, coordinated game plan was a sure guarantee of confusion and conflict.[96] In comparison, Genentech stood out as a nose-to-the-grindstone, eyes-fixed-on-the-goal, product-oriented operation—Swanson's basic business philosophy writ large.

Second only to his fixation on product focus was his insistence that Genentech achieve break-even earnings, if not profitability, even as a struggling start-up.[97] The company reached that goal in surprisingly short order. In April 1979, in what may have been Genentech's first full financial analysis, Fred Middleton reported that the company had become "a full-fledged 'revenue-producing' business."[98] "Full-fledged" was a dubious appellation for a company barely out of infancy. All the more striking, then, was Middleton's claim of financial solvency for a start-up but three years old. He asserted that, despite the company's rapid expansion and heavy research expenses, Genentech had been operating in the black since the third quarter of 1978 and had ended that year with a cash balance of $950,000. Swanson could now claim to potential investors, as he promptly proceeded to do, that the firm's financial condition was sound, its cash flow positive, and its revenues sufficient to cover operating expenses.[99] Genentech's financial solvency three years into its existence was for Swanson a point of pride and a sign of corporate discipline.[100]

Despite the fixity of Swanson's corporate goals and bulldog tenacity, his management style was conspicuously informal and interactive. In dealings with the scientists, he was mainly a facilitator and cheerleader. With no background in molecular biology except what he picked up, he could only reassure and applaud from the sidelines. Yet he delighted in popping unannounced into the labs, looking over the scientists' shoulders, asking questions, exulting over positive results, goading everyone on when results disappointed. It was a form of hands-on management that one of his heroes, David Packard of Hewlett-Packard renown (and a Genentech director as of 1981), called "management-by-walking-around."[101] Swanson would continue this highly personal practice well into the 1980s, until the company grew too large.

At executive board meetings and on periodic rounds of the labs, Boyer was also casually contributory and helpfully communicative. Swanson recalled:

> Herb was always interactive, primarily at the board level where the basic questions of which projects we should work on were decided. He had a clear insight of what the technical feasibility was, and where you couldn't push the science too far. Was it ready now? That was a big contribution. Boyer was always the one. I think his judgment calls were critical. . . .
>
> The other thing Herb did in those early days was wander around the labs, as I did. Where my job was "Okay, where are we on this [project]?" and to act as cheerleader to get people fired up and to coordinate between groups; his was, in a sense, [to act as] a scientific sounding board. "Okay, here's how I'm approaching this." So he was somebody to talk to about the scientific details. He did that very well.[102]

Although early on Genentech utilized an array of paid consultants, the fact remained that for many years there was no senior scientific authority regularly on site with the formal responsibility to direct and oversee research. As a result, the scientists had unusual latitude in making decisions about the conduct and direction of experiments. It was a new experience for many of them, accustomed to being under the thumb of the professor at the head of the lab. The three quick gene-cloning successes indicate that this latitude paid off in fostering ingenuity, productivity, and self-confidence in the firm's first generation of scientists. Only in 1983 did Genentech create the formal position of vice president of research and appoint a UCSF professor of molecular biology to fill it.

But the absence of a formal research director prior to that time had a possible downside. A director providing structured interaction between management and scientists might have allayed the early scientists' concern over lack of representation in corporate decision-making. Early in 1979 a worried Dennis Kleid rounded up the ten or so scientists with doctoral degrees to discuss the issue in private in an upstairs room of a shabby restaurant, which only added to the meeting's subversive feel.[103] After the meeting, Kleid asked Swanson to place at least one scientist on the management committee and one on the board of directors. Swanson immediately agreed to the former request but adamantly refused the latter, most likely because board members are traditionally outsiders.[104] The eventual result was greater transparency between management and research, although Swanson remained of a general mind that business was for businesspeople and scientists should stick to science.

AN EMERGING CULTURE

A culture was taking shape at Genentech that had no exact counterpart in industry or academia. The high-tech firms of Silicon Valley and along Route 128 in Massachusetts shared its emphasis on innovation, fast-moving research, and intellectual property creation and protection. But the electronics and computer industries, and every other industrial sector for that matter, lacked the close, significant, and sustained ties with university research that Genentech drew upon from the start and that continue to define the biotechnology industry of today. Virtually every element in the company's research endeavor—from its scientists to its intellectual and technological foundations—had originated in decade upon decade of accumulated basic-science knowledge generated in academic labs. Boyer and the early scientists were the medium, the very embodiment, of the knowledge and techniques upon which Genentech was built. Inevitably, along with the science and technology, came the cultural values, expectations, and conventions of academic life. At Boyer's insistence, the scientists were encouraged to publish and engage in the wider community of science. The policy resulted in collaborations and competition with university scientists and the kudos and status accruing from Genentech's research findings and mounting record of publications in major scientific journals.

But academic values had to accommodate corporate realities: at Swanson's insistence, research was to lead to strong patents, marketable products, and profit. Genentech's culture was in short a hybrid of academic values brought in line with commercial objectives and practices. It was, to turn a phrase, a "recombinant culture" in ways that the biotechnology industry of today continues to manifest in one way or another. As Swanson put it, "We have provided an academic atmosphere with industrial focus and resources."[105] His prosaic statement does not begin to capture the over-the-top dynamism of Genentech's early culture.

The environment of the novel enterprise in South San Francisco comprised a distinctive mix of scientific intensity, male adrenaline, and juvenile letting off steam. Visitors were immediately struck by the air of energy and electricity. The young scientists banded together into flexible multidisciplinary teams that exhibited inexhaustible engagement, camaraderie, and a willingness to pull together to reach common ends. Heyneker contrasted Genentech's evident teamwork with academia's bias toward rewarding individual endeavor:

> In academe, the motivation is quite different. Graduate students are there to get a PhD thesis, so they focus on their little aspect. That's all there is to it. They don't have to integrate into a bigger project. The postdocs are there to make a name for themselves because they want to become assistant professors, so they have to publish. Those are the most productive years. But again, the goal is very personal. "What contribution can I make to a certain understanding of whatever." It can be very individualistic.
>
> In industry, the goals are more clearly defined, but often you need different disciplines to reach them. So, indeed, out of Genentech came articles with twelve or fifteen names on them, and it was always viewed by academe as a funny way of doing science. I found the contrary; it was a very different way of doing science, because this was a demonstration that you can accomplish a lot by working together with different disciplines.[106]

Goeddel was an extreme expression of concentrated energy and unfaltering drive. But everyone reflected, in one way or another, a work ethic that had no tolerance for slackers or less than complete dedication and willingness to work long hours. Swanson was the supportive but insistent slave driver, urging on employees beyond their perceived limits. As Dan Yansura remembered:

> Bob wanted everything. He would say, "If you don't have more things on your plate than you can accomplish, then you're not trying hard enough." He wanted you to have a large enough list that you couldn't possibly get everything done, and yet he wanted you to try.[107]

Visitors were sometimes taken aback by the pervasive informality and irreverence for seniority. Touring the premises, they sometimes had to avoid clusters of scientists bowling for dollars in hallways or playing foosball while waiting for experiments to run their course. A delegation from Japan was visibly disconcerted when Swanson stopped in the midst of a facilities tour to fix a leaky water fountain.

Fledgling start-ups pitted against pharmaceutical giants could compete mainly by being more innovative, aggressive, and fleet of foot. Early Genentech had those attributes in spades. Swanson expected— demanded—a lot of everyone. His attitude was, as Roberto Crea recalled:

> Go get it; be there first; we have to beat everybody else. . . . We were small, undercapitalized, and relatively unknown to the world. We had to perform better than anybody else to gain legitimacy in the new industry. Once we did, we wanted to maintain the leadership.[108]

Venture capitalist Perkins, after a decade and more of working closely with Swanson, observed succinctly: "Bob would never be accused of lacking a sense of urgency."[109] The scientists, however, needed little prompting. From the start, they were motivated to show themselves equal, if not superior, to their academic peers, many of whom considered industrial scientists an inferior bunch. Crea again: "We as the scientists were very much concerned that the quality of the science at Genentech would be as high or higher than [that of] any academic laboratory, because we didn't want to be perceived as mercenaries or second-tier scientists."[110] Even Ullrich, despite European discomfort with raucous American behavior, admitted to being seduced by Genentech's unswervingly committed, can-do culture:

Even though I was not the prime example of such a [fervent] employee, I was just pulled in. We were very excited about Genentech and this feeling of belonging—very important. Rationally or not, I was just pulled into this stream of conviction that we were on the right way and that we were doing something important and exciting.[111]

The firm's "Ho-hos," all-company beer fests on Friday afternoons, provide a window on its exuberant early culture. The first Ho-ho, a notably modest affair, occurred in June 1978, a few days after the company's labs opened. All six employees, plus Kiley, attended, prewarned of the requirement to drink at least one glass of beer.[112] Kiley retains a mental image of CEO Swanson, salami and bread clutched in one hand and a glass of beer in the other, indistinguishable from the circle of young employees except for his lack of running shoes and jeans.[113] The scene was light-years distant from the top-down management and rigid hierarchy of established corporations with their firm divisions and sharp demarcation between management and staff. As the company grew, the Ho-hos remained pointedly egalitarian and for a time wildly raucous, untamed occasions where everyone socialized and decompressed.[114] Swanson, the management team, and sometimes Boyer, all tie-less and in shirtsleeves, mixed and kibitzed on a first-name basis. Kiley described one Ho-ho, which struck his eccentric fancy:

> When the right monkeys lived and the right monkeys died in an experiment involving interferon, Swanson and I declared the "Combined Simian Memorial and Revival Ho-ho" to which we wore gorilla outfits. I recall being in line at the cash register with Bob when we were shopping for the Ho-ho— bananas, peanuts, things of that nature—telling him that we had forgotten the paper plates, to which he replied, "Monkeys—don't—need—plates."[115]

By the late 1980s, Genentech had reined in the antics, assuming a modicum of the decorum appropriate for a maturing company.

The egalitarianism was even mapped onto the physical landscape: there were no reserved parking places and, as Genentech expanded, offices were designed to be of roughly the same size and appearance to avoid any visible sign of seniority. Even Swanson throughout his tenure as CEO—he left in 1990—kept his modest corner office in Building 1. The egalitarianism so evident at Genentech may have been at an extreme end of the spectrum, but it was not unique in the business world, particularly in Silicon Valley. The 1970s found many industries adopting nonhierarchical organization aimed at greater employee participation and dedication to the firm.[116] But in comparison to the average pharmaceutical company, Genentech's culture was stunningly new. Its flat organization, turn-on-a-dime flexibility, and cocky, no-holds-barred inventiveness were a study in contrast to Big Pharma's executive-lunchroom culture, measured pace, and stodgy regimentation.

But in a culture of the young, the spirited, the ambitious, and the irreverent, boundaries were more than occasionally crossed and proprieties ignored. Pranks and high jinks released tensions bred in the hothouse climate of science-to-win. The atmosphere was overwhelmingly, aggressively, inescapably masculine.[117] A female scientist, a postdoc at Genentech in the early 1980s, observed: "The company seemed to operate like a boys' locker room, and the place reeked of testosterone. No prank was too outrageous, no poker bet was too high, and no woman was part of the inner circle."[118] It was not only women who noticed the heavily masculine culture. Laurence Lasky, a molecular biologist who arrived at Genentech in 1982, commented that the company was "macho city," a place with pinups on the wall and "no thought police."[119] Reflecting the cultural norms of the 1970s, a decade less socially conscious and inclusive than a later era, Genentech's culture of extremes included a strand that observers today would label socially unacceptable. But it was not Genentech's blemishes that financiers noticed. They saw a company with an impressive line of scientific accomplishments and major corporate alliances. More and more, as 1980 approached, they wanted a piece of biotech action.

6

Wall Street Debut

"Genetic Firm's Stock Starts Wall St. Frenzy."
Los Angeles Times, October 15, 1980[1]

BIOMANIA

As the 1970s drew to a close, public appetite skyrocketed for news of the latest genetic engineering accomplishments. The media enthusiastically—and often uncritically—fanned the flames, reporting what wonders recombinant DNA might accomplish in making a cornucopia of new products in the medical, energy, and nutrition fields. In one of many possible examples, a *Wall Street Journal* article conveyed the impression that biotechnology was about to bear abundant fruit:

> At the beginning of the 1970s, it sounded like the further reaches of science fiction: turning microbes into factories of "superbugs" producing everything from food to energy to medicine. Now, judging by the publicity emerging from a number of companies in the field, it is fast approaching reality.[2]

In November 1979 an article in *Science*, evocatively titled, "Recombinant DNA: Warming Up for Big Payoff," reported on the influx of money from giant multinationals and financial institutions into the scattering of genetic engineering companies created in the late 1970s.[3] The previous August, an investment analyst at E. F. Hutton had arranged a seminar on biotechnology featuring talks by executives of Genentech, Cetus, Biogen, and Genex. Instead of the expected thirty-five participants, the conference attracted five hundred, mainly institutional investors interested in learning about this latest technological phenomenon.[4] The unalloyed message was that big money awaited to be made. Cetus's and Genentech's array of alliances with major corporations reinforced that assumption.[5] By late summer of 1979, venture capital and corporate investment in industrial biology totaled an estimated $150 million.[6] A year or so later, a government report calculated that one hundred U.S. companies were currently conducting or evaluating recombinant DNA or other tools of the new biology.[7] "Biomania," a term coined at this time, was sweeping the financial and corporate worlds.

A shift in the national environment for high technology and small business fueled the craze. A rising chorus complained of "outmoded patent laws, restrictive taxes and onerous regulations" stifling American ingenuity and willingness to take investment risks in fields like fiber optics, semiconductors, computers, and biotechnology.[8] Policy makers came to recognize that stiff environmental and health and safety regulations hampered technological development and commercial exploitation of basic-science discoveries.[9] During the Carter presidency of the late 1970s and continuing full force under Reagan in the 1980s, Congress, in a sweeping change of stance, passed a number of pro-business, pro-technology initiatives aimed at stimulating a sagging national economy and fostering international competitiveness. In 1978, in an effort to encourage investment, Congress had cut the tax on long-term capital gains, and a year later relaxed the so-called prudent man rules restricting pension fund investment in high-risk/high-return endeavors. The new legislation prompted investors to move out of tax shelters and into the stock market, where the hunt was on for attractive investment opportunities. In 1980 Congress followed suit with the Stevenson-Wydler Technology Innovation Act and the Bayh-Dole Patent and Trademark Act, both designed to ease and encourage patenting and licensing of the results of federally funded research and in so doing foster technology transfer to the private sector and stimulate U.S. productivity. How responsible such legislation was for sparking commercialization is debated.[10] The history recounted here documents a rising preoccupation with technology transfer and patenting at Stanford, UC, and Harvard that preceded Bayh-Dole by several years. At the very least, passage of pro-business legislation around 1980 indicated a decided turnabout in Washington policy from earlier preoccupation with technological danger and risk toward promotion of industrial productivity and research-based businesses as vehicles for helping to restore the nation to its rightful position as a world leader in high technology.

National attitudes toward recombinant DNA research were likewise shifting. Earlier concerns about health and environmental safety, although not entirely absent from public debate, were giving way to expectations for the technology paying off in novel products in major industrial areas. Visions of the myriad practical benefits stemming from the new genetic technologies were fast overcoming previous worry over possible biohazards. Prominent scientists drummed home the message that a half dozen years of genetic engineering research had demonstrated no clear evidence of danger to human health or the natural environment.[11] The benefits, the argument went, far outshone the risks. In a nation bent on preserving its leadership in the new genetics, the few biotech companies existing around 1980 enjoyed greater regulatory and bureaucratic latitude: the threat of federal legislation restricting recombinant DNA research had passed, the revised NIH guidelines were notably more lenient, and the Recombinant DNA Advisory Committee had made several concessions to industry, including holding secret sessions to protect proprietary information and permitting large-volume, manufacturing-size cultures of recombinant organisms.[12] Biotechnology firms counted prominently among the centers of American high technology that Washington now favored in its strategy for restoring the United States to an international position of innovative power and competitiveness. It was in this propitious new environment for high-tech business that Genentech began to consider a consequential change in corporate status.

EXIT STRATEGIES

Creating human insulin and human growth hormone had indicated the power and scope of Genentech's technology. But a company substantially supported by venture capital needed more than technological achievement; it needed to provide financial return to its investors, and sooner rather than later. In committing venture funds to Genentech, Eugene Kleiner and Tom Perkins had assumed that within a few years either an established corporation would acquire the company or, better yet,

Genentech would stage a public stock offering. Through one or the other of these "exit strategies," as venture capitalists called them, Kleiner & Perkins and its co-investors could "cash in," and in so doing fulfill their primary responsibility: to recoup for their fund investors and for themselves their original investment, hopefully at many times the initial stake.[13] Thus, it was never a question *if* Genentech would take one or the other exit strategy; the question was solely *when*. As chairman of the board of directors, Perkins had considerable power in deciding such matters. In light of the glowing media attention to the insulin and growth hormone successes and the firm's impressive corporate alliances, he concluded that Genentech's moment had come. He advocated a buyout by an established corporation and, with Swanson and others, approached Johnson & Johnson, the venerable maker of Band-Aids and other familiar items in American medicine cabinets.

Hoping to move negotiations along, Perkins hosted a dinner party for the J&J president and other executives at his elegant Marin County, California, estate. With its sweeping views of San Francisco Bay, a garage full of restored vintage automobiles, and an elegant yacht moored in nearby Sausalito, the occasion—complete with formal attire and a staff of uniformed servants—was clearly meant to impress. And impress it did, particularly Swanson, Boyer, and one or two Genentech managers, all feeling somewhat intimidated in the rarefied surroundings. Goeddel showed up late, coming straight from a long Saturday of research at Genentech. Dressed in his usual jeans, T-shirt, and running shoes, he had trouble getting past the tuxedoed butler until Perkins gave word to let him in. Feeling like a fish out of water, Goeddel apologized to Swanson for his informal attire. Swanson told him privately with obvious delight, "You look like a scientist. This is great"—a scientist so engaged in his experiments that he could not spare time to dress up.[14] Imposing though the occasion was, it did not dispel J&J's qualms about acquiring an unorthodox company with a strange technology. Testing the wind at a later meeting, Perkins floated the idea of a purchase price of $80 million. The offer fell flat. Fred Middleton, present at the negotiations, speculated that J&J didn't have "a clue as to what to do with this [recombinant DNA] technology—certainly didn't know what it was worth. They couldn't fit it into a Band-Aid mold."[15] Flummoxed by a new kind of company with an unfamiliar approach to making drugs, J&J executives were unsure how to value Genentech, there being no standard for comparison or history of earnings. They doubtless also worried about the lack of immediate products and the difficulty of integrating a technology involving large biological molecules into traditional small-molecule pharmaceutical manufacturing practice.

Perkins and Swanson made one more attempt to sell Genentech. Late in 1979 Perkins, Swanson, Kiley, and Middleton boarded a plane for Indianapolis to meet with Eli Lilly's CEO and others in top management. Figuring that Genentech's insulin and growth hormone clones were its aces in the hole, Perkins suggested a selling price of $100 million.[16] After several subsequent meetings, the negotiations petered out and any chance of a deal fell through. Middleton's view is that Lilly was hamstrung by a conservative "not invented here" mentality, an opinion supported by the drug firm's reputation for relying primarily on internal research and only reluctantly entering into outside contracts.[17] But it might also be that Lilly for the moment had all that it wanted from the upstart firm—the human insulin-producing clones. Whatever the case, one possible route of Genentech's development—acquisition—was closed for the foreseeable future.

After the failure to find a buyer, Genentech's directors concluded that the company's technology was too novel, too experimental, too unconventional for a conservative pharmaceutical industry to adopt wholeheartedly. Perkins then broached the idea of a public stock offering in which the firm would offer shares for purchase on the stock market. An offering, Perkins forcefully insisted, was a means to raise money—perhaps *big* money—more than was generally available through venture-capital financing. Furthermore, he pointed out, the stock market was finally turning favorable to initial public offerings.[18] The market, for most of the 1970s in the doldrums, was on the rise in 1980, and financiers, encouraged by the newly favorable tax and investment laws, were eager to invest. A result was that in 1980 the total amount of capital invested in new business ventures rose an estimated 50 percent over the previous year.[19]

But it was not only the increasingly auspicious economic climate that prompted Perkins to advocate an IPO. He regarded Genentech's interferon research, with its projected billion-dollar market, to be a major asset in bringing public attention to the company and in endowing its march to the public financial markets with the aura of miracle cures and big money.

By the late 1970s, anyone following commercial biotechnology was entranced by news about interferon, a protein discovered in 1957 and thought to prevent virus infection.[20] Interest in the mysterious substance soared in the 1970s during the Nixon administration's War on Cancer with its underlying assumption that viruses caused many types of tumors. A prevailing expectation was that if interferon suppressed viral infections, it should prove effective in fighting cancer. The Finnish virologist Kari Cantell had devoted a professional lifetime to devising methods for laboriously extracting minute quantities of natural interferon from human white blood cells grown in cell culture.[21] The Finnish Red Cross adopted his methods and became the main distributor of the precious but impure and exorbitantly costly substance. In the late 1970s, the marginally favorable results of clinical trials of the Finnish interferon inflated public expectations for a cancer cure and stoked an intensely competitive international effort involving dozens of academic and industrial laboratories pursuing various approaches to interferon production. "The drug companies know that there is a gold mine in interferon," *Time* magazine reported in a March 1980 cover story. "They are scrambling like mad to produce it."[22] *Time*'s overwrought coverage reflected the hype the media brought to a suspected cancer remedy. Although it was clear by this time that three major subtypes of interferon existed—fibroblast, leukocyte, and immune interferon—precisely how they worked and what therapeutic value they had, if any, remained clinically unsubstantiated.[23]

Pharmaceuticals with the aura of immense market potential but handicapped by impurity, minuscule quantity, and high cost were obvious candidates for genetic engineering. Biogen's formation in 1978 was in considerable part premised on recombinant interferon research, and investigators at Cetus, DuPont, Hoffmann–La Roche, Harvard, Caltech, and elsewhere were also competing to clone the interferon genes—regarded at the time as the ultimate challenge and high-stakes payoff of recombinant DNA research. From the start, Genentech had considered interferon a possible research target but had resisted jumping on the bandwagon until it had made better-understood proteins with existing markets. Insulin and growth hormone were the low-hanging fruit, as the scientists referred to them, recombinant replacements for natural hormones with well-documented medical uses and public demand.

Interferon, whose protein sequence was unknown and for which no established market existed, was in a far riskier category and hence, Genentech's scientist deemed, an inappropriate initial target. Nonetheless, in November 1978 Swanson, never loath to push the envelope and eager to conclude research and development agreements, signed a confidential letter of intent with the Swiss pharmaceutical firm Hoffmann–La Roche to develop bacteria producing interferon.[24] Confident as a result of the growth hormone work that it could make complex proteins in quantity, Genentech was ready by mid-1979 to launch research in riskier directions. The firm was set to develop, as Swanson announced to stockbrokers that fall, "new products . . . for entirely new markets."[25] One of those products was to be recombinant interferon.

On January 6, 1980, Hoffmann–La Roche and Genentech signed a formal agreement to collaborate in research on leukocyte and fibroblast interferon. Both parties were to develop and manufacture products, but Roche alone would be responsible for marketing. Aside from the requisite financial support, Genentech needed the experience and clout of a major pharmaceutical company to develop a new and purportedly enormous market for the interferons—if indeed their many predicted indications materialized. But Roche provided more than money and experience. Sidney Pestka, a respected protein biochemist working on natural interferon at Roche's New Jersey research institute, possessed several interferon-producing cell lines, invaluable sources of the messenger RNA that Genentech would need to make complementary DNA copies of interferon genes.

Genentech had just signed the contract when the news broke, on January 16, that Charles Weissmann's lab at the University of Geneva, in a project Biogen supported, had cloned and expressed a precursor of leukocyte interferon.[26] Indicating the hot commercial prospects expected for interferon, a Biogen attorney had flown to Geneva at Christmastime to meet with Weissmann and file a patent application a week before the announcement. By fortunate happenstance, Genentech had concluded its contract with Roche in the nick of time. As Kiley recalled, "Had [Biogen's] announcement come just a week sooner, I doubt very much we would have made our deal with Roche."[27] At a splashy press conference, Weissmann and Harvard's Wally Gilbert, both Biogen directors, went all out to promote the company's achievement, describing interferon as "a protein of dramatic medical interest."[28] The synthesis of an interferon with its mystique of a cancer cure prompted worldwide headlines and drew further attention to biotechnology as an exhilarating high-tech arena. *Fortune* magazine commented that Weissmann's interferon discovery "added a lot of fuel to the firestorm of publicity about genetic engineering."[29]

But the elation was not universal. Critics immediately attacked Weissmann for performing commercial research in his university lab—a reprise of Boyer's experience with somatostatin. Observers also noticed that Biogen's rushed announcement—its directors feared Genentech would scoop them—came before the Weissmann lab had determined the sequence of its interferon. Clearly, the biomedical research establishment had not come to terms with the commercialism increasingly apparent in its ranks. Moreover, the previous year a Japanese team had identified and published a partial sequence of an interferon gene.[30] As the *Science* reporter Nicholas Wade tartly remarked regarding Biogen's fanfare, "A major announcement in molecular biology this was not."[31] Yet to the financial community, the questionable scientific and clinical significance of the discovery seemed scarcely to matter. Wade's article, with its telling title, "Cloning Gold Rush Turns Basic Biology into Big Business," conveyed the overblown reaction to the medical and financial possibilities of interferon and other pharmaceuticals touted as within biotechnology's reach.

Unfazed by Biogen's announcement, Goeddel and colleagues focused on cloning and expressing interferon, refusing to let the stiff competition intimidate or distract them. For Goeddel, the very competitiveness and import of the project had immense appeal. Interferon was, as he later put it, "the 'sexiest' project possible for cloning, and I'm sure from a commercial point of view the one with the largest potential payoff."[32] Achieving interferon would mean more to the company than celebrated scientific achievement and expected market bonanza. Making interferon with its projected billion-dollar market would provide a launching pad for Genentech to go public. Much obviously rode on the project's success.

However, making recombinant interferon presented exceptional problems. Goeddel recalled discussions characterizing it as "a kind of magical substance" whose biological mechanism was unknown and whose very existence was in question.[33] An immediate practical problem was that interferon's complete protein sequence remained unknown, making it problematic for Crea to construct the synthetic DNA probes necessary to identify the interferon sequence. Then, in a stroke of good fortune, Heyneker and a colleague attended a scientific meeting in which the speaker—to everyone's astonishment given the field's intense competitiveness—projected a slide of a partial protein sequence of fibroblast interferon. Elatedly jotting down the chemical structure, they telephoned the information to Goeddel, who instantly relayed the sequence order to Crea. With the invaluable information, Crea immediately started to construct the required probes. The Genentech scientists now had a means to fish out the interferon messenger RNA for constructing the complementary DNA sequence of the gene itself. But it took day after day of tedious screening to find the interferon messenger RNA in the welter of biological material. In another round of onerous labor, Goeddel constructed a "library" of thousands upon thousands of bacterial cells containing a multitude of complementary DNA sequences. He and his colleagues then painstakingly searched through the cells, seeking ones with the interferon gene. Pestka, it happened, was the first to identify a bacterial colony thought to contain interferon. Using the partial sequence Pestka retrieved, Goeddel cloned full-length DNA sequences for both fibroblast and leukocyte interferon and within weeks had expressed the two types of interferon.[34]

In June 1980, after filing for patent protection, Genentech announced production in collaboration with Roche of the two interferons, noting reports of their potential activity against tumors and viruses.[35] In gibes aimed at Biogen, the announcement made much of the fact that the collaboration had produced complete, active, and pure forms of the interferons and in higher yields than anyone had previously reported. Genentech had, in fact, made enough of the rare substances to inject into three monkeys and claim protection against virus challenge. The firm's caution that sufficient quantity of pure interferon would not be available for clinical trials in humans until 1981 only whetted public expectation of an imminent cancer remedy. Mindful of that possibility, the NIH Recombinant DNA Advisory Committee (RAC), meeting a few days after the announcement in closed session (a concession to industry demands to protect corporate secrets), recommended approving Genentech's application to produce interferon in 600-liter batches, multiple times the original ten-liter limit. The New York Times noted that the RAC's decision would allow the firm to manufacture more interferon in a week than conventional means could produce in a year.[36] Rumors of an imminent Genentech public offering immediately soared.[37]

Following close on the heels of the growth hormone achievement, the firm staged another raucous celebration. Jubilant partygoers costumed as monkeys—reminders of those used in the pre-clinical trials of interferon—brandished bananas to toast another triumph. But despite high expectations and inflated predictions, the interferons did not for some time live up to the forecasts of miracle cures and financial windfall. Only after years of research—and discovery of a large family of distinct interferon proteins—were some transformed from orphan drugs into FDA-approved therapy for certain kinds of cancer, multiple sclerosis, and other diseases.[38] In 2005 interferon products garnered for their various manufacturers an estimated $5 billion in global sales.[39]

Intoxicated by Genentech's and Biogen's interferon successes, the financial sector's ardor for biotechnology approached a peak in 1980. If biotechnology could produce substances as rare and valuable as the interferons, there seemed no limit to its commercial potential. *Fortune* magazine reported in June that the paper value of the four leading biotech companies—Cetus, Genentech, Genex, and Biogen—had doubled in six months to a total of $500 million. The article went on to comment that "recombinant DNA has suddenly emerged as one of the hottest investment fields of the new decade."[40] The first issue of *BioEngineering News*—published in autumn 1980 as the self-proclaimed "newsletter of the genetic engineering industry"—stated that an estimated $1 billion had been invested in commercial development of recombinant DNA technology.[41] An E. F. Hutton stock market analyst prone to over-promotion of biotechnology as an investment area remarked: "You can just feel the excitement in the air. Here we are sitting at the edge of a technological breakthrough that could be as important as electricity, splitting the atom, or going back to the invention of the wheel or discovery of fire."[42] Biotechnology was fast becoming the new glamour sector, closely tracked by financiers and investors willing to take risks for anticipated high returns.

RUN-UP TO AN INITIAL PUBLIC OFFERING

By common standards of conventional business practice, Perkins was jumping the gun in pushing for Genentech to go public. Companies at the time usually waited until they had one or more products on or very near the market and at least a semblance of sales revenues before staging a public offering.[43] Genentech of course had neither. If the firm went forward with a stock offering, it meant asking for public investment earlier in the corporate life cycle than was customary. Yet Perkins believed that Genentech had to go sooner than later to the public markets, where money was sufficient to support corporate development and to finance costly clinical trials. Swanson, he argued, could not fulfill his dream of a fully integrated, independent Genentech by limping along on periodic risk-capital infusions and corporate benchmark payments. And then there was Perkins's compelling obligation to satisfy his venture-fund investors by a return in cold, hard cash. He also saw robust competitive advantages for an immediate IPO, as he recalled years later:

> I was very keen on taking Genentech public because I thought—I was right—we would dominate the field, we'd suck up all the oxygen, and we would be able to use our new celebrity status to hire the best people in the world to help us grow. All of which we did. *And* we could use our public price to set the stage for subsequent rounds of financing, which we did do and which we had to have, in [negotiations in] Japan and Europe and all kinds of places.[44]

But public market financing would come at a price—rigorous Securities and Exchange Commission (SEC) reporting requirements, stockholder and analyst scrutiny, and public pressure to achieve milestones and reach profitability. Swanson stubbornly resisted Perkins's proposal, leery of the early timing and resistant to having to manage a company in the full glare of public and regulatory surveillance. But Swanson's most strident objection was to revealing sensitive product and contractual information to competitors in the detailed prospectus for general distribution that the SEC required of companies going public. The dispute became a standoff between the forceful venture capitalist and the obdurate thirty-two-year-old. As Perkins recalled:

> The only major quarrel [Bob and I] ever had was over when to take the company public. I felt very strongly that we should be the first, that it would nail down Genentech's position as the leader. It would be horrible if a flaky outfit like Cetus were to be first. Swanson and I really quarreled about this. He knew that we needed to be public, but it was a huge amount of additional work for him. My telephone wasn't going to ring from irate investment analysts and shareholders. His was, and he knew that.[45]

Perkins persisted in campaigning forcefully for executive board agreement to stage an IPO. "I want you to know," he told board members, "that the [market] timing is perfect. All the planets are lined up. Are we ready to go on this?"[46] He was adamant that Genentech be the first biotech company to reach Wall Street and reap the considerable public-relations advantage of being the front-runner in a sizzling industrial area. Swanson's continued resistance pushed an irked Perkins to demand a vote. With only three board members present—Boyer, Perkins, and Swanson (Lubrizol's Donald Murfin was absent)—Boyer's would be the deciding vote. Perkins and Swanson, in heated argument, put Boyer on the spot, asking how he intended to vote. Hesitating a moment, Boyer replied, "I always vote with my friends." His wily answer broke the ice— everyone laughed.[47] Over lunch with Swanson the next day, Perkins continued his campaign. Playing on Swanson's strong competitive streak, above all with Cetus, Perkins again raised the specter of a rival beating Genentech to Wall Street. That did it, according to Perkins; he convinced Swanson to commit to an IPO.[48] Yet even with Swanson on board, the way was not clear. Two legal issues loomed large and forbidding, threatening the offering.

LEGAL IMPEDIMENTS

As was everyone associated with commercial biotechnology, Swanson was concerned about the outcome of a high-profile lawsuit. The *Diamond v. Chakrabarty* case on the patentability of living organisms was on the Supreme Court's docket for argument in the spring of 1980.[49] The specific legal question concerned whether General Electric could patent a bacterium that Ananda Chakrabarty, a GE biochemist, had constructed by non-recombinant means to degrade crude oil. The commissioner at the U.S. Patent and Trademark Office had rejected GE's application, arguing that the patent office could not issue a patent on a living microorganism, considered a product of nature and therefore not patentable under the U.S. Constitution. The commissioner then refused to allow the patent office to review all patent applications involving living organisms or their components until the courts arrived at a decision. The result was a backlog in 1980 of a hundred or more applications, Genentech's included.[50]

Until the patentability issue was definitively resolved, Genentech and other companies dealing with living organisms or their components had to rely on trade secrecy or patent applications on the product and/or process, rather than on the engineered organism itself. Noting the importance of *Chakrabarty* to young biotech firms desperate for investment, the *Wall Street Journal* observed, "The pending patent applications [in biotechnology] have become a central issue that has ramifications beyond research. The small companies littering the field need the exclusivity that patents afford to attract the capital they need to expand."[51] Swanson fretted that a ruling forbidding the patenting of life forms might derail a stock offering by deflecting potential investors, dubious of Genentech's ability to legally protect its living inventions. Asked in 1996 if he had worried about the Supreme Court's decision, Swanson replied:

> I was worried about a lot of things. I was worried most of the time. But you sort of put a brave face on it and say, well, maybe if the organism itself isn't patentable, maybe some of the genes that you've made are patentable. Obviously, the more you can wrap patents around what you're doing, the better off you are.[52]

It fell to Tom Kiley to write a friend-of-the-court brief for the Supreme Court's deliberation of the *Chakrabarty* case. It was among the nine briefs submitted, all but one from institutions with an economic stake in the patentability of living organisms. Kiley's brief, listing Genentech's impressive run of gene-cloning successes—each pointedly noted as buttressed by patent applications—underlined the importance of intellectual property protection in attracting investment in a pathbreaking area of biomedical research. "In Genentech's case," he wrote, "the patent incentive did, and doubtless elsewhere it will, prove to be an important factor in attracting support for life-giving research."[53] In time-honored fashion, he was playing upon the justices' presumed sensitivity to the economic and social consequences of their decisions. He concluded by highlighting an issue on many minds, the nation's declining international leadership in high technology:

> The encouragement of domestic innovation is important, and that can best be done by a strengthened patent system, as both Congress and the President have agreed. In the important field of genetic engineering, that system would be best strengthened . . . by the grant of patents on microorganisms.[54]

Kiley had shrewdly linked the patenting of living things to the resumption of the nation's proper place at the international forefront of technological invention.

In June 1980 the Supreme Court ruled in a five-to-four decision that Chakrabarty's organism was not a product of nature but instead a novel invention of his own ingenuity and hence patentable under the Constitution as a new composition of matter. The justices decided in a narrow statutory interpretation, deliberately eschewing consideration of related moral arguments, that the distinction was not between living and inanimate things, but rather between products of nature, whether living or not, and human-made inventions. To buttress its decision, the Court cited a memorable line in a Senate report maintaining that "anything under the sun made by man is patentable subject matter."[55] The justices' ruling removed any inherent differentiation in U.S. patent law between living and nonliving matter and opened the biological world to the possibility of proprietary ownership and subsequent commercialization. The patent system henceforth could be—and indeed soon was—used for securing private ownership of all manner of living organisms and their components.[56] As of December 1980, the Patent Office was processing approximately two hundred applications on microorganisms.[57] The Supreme Court decision in *Chakrabarty* would become a cornerstone of biotechnology law and a significant milestone in the commercialization of biology.

Jeremy Rifkin and his organization, the People's Business Commission—which had submitted the sole brief arguing against the patenting of life forms—were appalled. Their brief presented the patenting of life as diametrically opposed to the public interest and argued that Congress never meant living things, engineered or not, to be patented.[58] Religious leaders around the world viewed the ruling as a terrifying incursion of commercial interests into the natural world of heavenly creation and a denial of life's sacred property. Predictably, Genentech heralded the decision as a positive sign for biotechnology. A corporate news release quoted an expansive Swanson: "The Court's decision should accelerate the flow of investment capital into new, high technology ventures. . . . By extending the reach of patents to encompass more than merely 'traditional' fields of research, the Court has assured this country's technology future."[59] Out of the public spotlight, Swanson and colleagues rejoiced that the decision had removed a potentially formidable legal and psychological roadblock to Genentech's IPO. Looking back on the memorable decision, Kiley remarked:

> When the Court in *Chakrabarty* said you could patent the microbes themselves, why, that was a famous decision—famous because it intrigued the public, amazed the public—patents on new life forms! And so it turned a spotlight on the industry. The decision was regarded as positive for the industry and undoubtedly was a boon to the public offering. Had the case gone the other way, one supposes the emotional reaction would have been very negative, and it might have been quite difficult for Genentech to go out [with an IPO] and for others to follow.[60]

The decision had an immediate effect. Corporations that had previously held back because of patenting uncertainties now began to leap in, adopting genetic engineering techniques in one form or another. As *Business Week* commented: "Now that the Supreme Court has cleared the way for patenting life forms made in the lab, virtually every big drug company, along with such disparate companies as Revlon, Brunswick, and Johnson & Johnson, are [*sic*] stepping up research in biotechnology."[61] A preoccupation of a financial sector avid to invest in biotechnology became how to acquire equity in the privately held start-ups, now swelled by several new entrants.[62] "Wall Street brokers," *Business Week* observed, "are beating the bushes for ways customers can get a piece of the action."[63] Despite the white-hot market for biotech issues, a troublesome legal obstacle to taking Genentech public remained.

Ullrich's and Seeburg's transfer of research materials to Genentech loomed as a formidable threat. Legal action by the University of California appeared to be a real possibility. In February 1980 Baxter, Goodman, and Rutter wrote to the UCSF chancellor accusing the university attorneys and its patent and licensing office of failing to pursue the matter of the UCSF material at Genentech with sufficient zeal. They asked for "aggressive action on the part of the University to force Genentech to compensate [the university] for these unfair acts" and to establish "a clear policy concerning the removal [from the university] of materials that are of commercial use."[64] Under threat of legal action, Genentech had to resolve the issue quickly. A legal confrontation with a major university, questioning the company's ownership of key research materials and shaking investors' confidence, was the last thing management wanted in the run-up to the IPO. "These are not circumstances," Kiley remarked in a deliberate understatement, "in which one expects a good reception on Wall Street."[65] In June 1980 Kiley met with a university attorney and proposed that Genentech make a cash payment of $250,000 to settle university grievances regarding the unauthorized transfer of insulin and growth hormone materials. Subsequent negotiations raised the payment to $350,000. Genentech's first annual report as a public company obliquely described "a one-time payment of $350,000 to the Regents of the University of California in connection with a restructuring of the Company's obligations arising from funded research."[66] The settlement avoided an immediate lawsuit that would have set back, if not destroyed, Genentech's plan for a public offering. The company had escaped disaster by the skin of its teeth. But the underlying legal issues would fester for years.

The long-standing problems of legal ownership of research materials and patent validity finally boiled to the surface in 1999 when the University of California's suit against Genentech for infringing its growth hormone patent came to trial.[67] The case, as a journalist put it, "rocked the biotech world."[68] A prominent university had sued a leading biotech company, revealing unsavory rivalries among former partners for intellectual property rights and major financial reward. At a spectacular turning point in the legal proceedings, Seeburg, by then an acting director of the Max Planck Institute for Medical Research in Heidelberg, Germany, reversed his sworn testimony in previous cases. He now claimed under oath that in 1978 he and Goeddel had used the complementary DNA clone he and John Shine had made at UCSF to express Genentech's human growth hormone.[69] Seeburg also testified that he and Goeddel had made a secret pact to conceal their use of the university material and that he and coauthors falsified technical data in Genentech's 1979 *Nature* paper to hide the UCSF origin of the complementary DNA clone.[70] Why, a university attorney queried, had he and Goeddel decided on secrecy? Seeburg replied: "We felt embarrassed that we couldn't make [the growth hormone experiment] work according to our plan; and then the other reason was that we didn't want . . . anyone else to know at U.C., for instance, because it might get us into problems."[71]

Goeddel testified vehemently and repeatedly that he and Seeburg never used the complementary DNA made at UCSF in Genentech's growth hormone project and flatly denied any secret pact.[72] Witnesses for the company corroborated Goeddel's testimony and claimed that lab notebooks showed the independence of the company's research. A Genentech attorney fiercely attacked Seeburg's credibility and maintained that the personal fortune in legal damages likely to ensue to him (and the other inventors on the UCSF patent) if the university won the case motivated his reversal in testimony. Eight Genentech scientists in subsequent letters published in *Science* and *Nature* categorically denied Seeburg's accusation that the 1979 growth hormone paper contained falsified data.[73] Seeburg's letter, published in the two journals alongside Genentech's, repeated his charge that a plasmid he and Shine constructed at UCSF was the source of the DNA in Genentech's growth hormone experiment.[74] The jury was unable to reach a verdict; the court scheduled a second trial for January 2000. In the interim, the two parties came to a settlement in 1999 that ended the drawn-out and exorbitantly expensive legal proceedings. Genentech agreed to pay the university $150 million and to make a $50 million contribution toward construction of a research building at UCSF's new Mission Bay campus in San Francisco.[75] The building, christened Genentech Hall, stands today at the center of campus, a symbolic reconciliation—or so both sides pointedly portrayed—of two long-term protagonists of biopharmaceutical research.[76]

In 1980 all this bitter litigation was of course in the unforeseeable future. What Genentech's board saw at the time was a clear path to a public stock offering. It was then up to Swanson, Middleton, Kiley, and a bevy of outside attorneys and underwriters to prepare the preliminary prospectus for the IPO.[77] In June the group began the complicated task of composing a document that had to describe the company's business, serve Genentech as a stock marketing tool, and also follow exacting SEC disclosure requirements. Genentech's prospectus was a particular challenge to write: no useful model existed, and rumor was that the SEC would focus close scrutiny on a public offering in a virgin field. Enzo Biochem, a company producing restriction enzymes and DNA probes, went public that June, with an offering that quickly sold out.[78] But Enzo's prospectus was of little help, attorneys for Genentech concluded, inadequately describing the emerging biotechnology industry and its risks. With no model to follow, no protocol for due diligence, no industry standards to guide them, the group quibbled among themselves and with the SEC over which risks it should disclose and how much it had to reveal of Genentech's heavily guarded contracts. The preliminary prospectus at last completed and submitted to the SEC, on August 19 Genentech issued a brief press release announcing registration of its securities with the SEC, a mandatory preliminary to a public offering.[79]

The preliminary prospectus anticipated an offering in September 1980 of one million shares, priced between $25 and $30 per share—a high price given Genentech's immaturity and notable lack of products and sales. The prospectus disclosed that the firm in the first six months of 1980 had earned $81,000 on revenues of $3.5 million, a disheartening price-to-earnings ratio. As Nicholas Wade bitingly observed in a *Science* editorial, "The company is not yet a gold mine."[80] Fred Middleton inadvertently agreed: "I think it's fair to say," he told a journalist, "that the value of our company is based on its future potential."[81] Promise and possibility, rather than products and earnings, fueled the speculative momentum building over the Genentech offering.

Wade's skepticism was not the rule; the financial world was ecstatic. Middleton spent his days fielding telephone calls from the press and panting investors, all wanting more details of the public offering. He tried to abide by SEC dictum that all Genentech personnel were to restrict public comments to information contained in the prospectus. Yet Middleton agreed—naively as he later admitted—to an interview with a local newspaper. He failed to anticipate that the reporter would do additional sleuthing and reiterate in the article a statement falsely attributed to Swanson that Genentech's goals were to have revenues of $100 million in the late 1980s and a staff of one thousand.[82] The SEC was incensed and placed an immediate hold on the pending IPO. Publication of such speculations was a clear violation of the SEC-mandated "quiet period" before and immediately after a public offering during which a company is to say nothing publicly that could influence the value of its stock. Middleton was aghast, panicked that Genentech's IPO would be indefinitely delayed as the SEC waited for the publicity's effects to fade. "That's what got us into hot water with the SEC—the so-called gun-jumping claim," Middleton recalled. "Because of all the stories in the media, the SEC thought we were out there actively hyping the deal."[83] The outside attorneys sprang into action, one in particular playing on his close ties with SEC officials. After a rough period in which the IPO came close to disintegrating, their arguments had the desired effect: the SEC relented, lifted the hold, and allowed Genentech's offering to proceed.[84]

On October 8 the SEC approved a share price of $35, a surprising $5–$10 higher than the previous range, and set an IPO date of October 14, 1980.[85]

Conforming to SEC requirements, the final prospectus made Genentech's high degree of risk abundantly clear. In regard to intellectual property protection, it was equally dour, reflecting current qualms about the unsettled status of patent rights in biotechnology, above and beyond the decision in *Chakrabarty*. Genentech expected to be issued patents, the prospectus stated, but cautioned that "there can be no assurance as to the breadth or the degree of protection which these patents, if issued, will afford the Company."[86] Furthermore, with universities also filing for patents in biotechnology, Genentech faced unknown fees if it had to license outside patents. The numerous risks added up, the prospectus warned, to a "highly volatile" share price.[87] *Caveat emptor*—let the buyer beware—was the message on every page.

Four pages of glossy colored photographs and a diagram inserted midway in the prospectus belied the warnings. A sophisticated two-page diagram under the heading "The Product Development Process" began with sketches of research materials, led the reader step-by-step through images of pharmaceutical development and manufacturing stages, and ended by picturing a physician and nurse standing by a patient's bedside over the caption "Health Care Applications." The last page in the insert, titled "Manufacturing Process," depicted a scientist dwarfed by a huge biopharmaceutical fermenter and a final shot of "the Company's products"—six labeled bottles partly filled with white substances and labeled Human Growth Hormone, Human Leukocyte Interferon, Human Proinsulin, and so on.[88] Here for the public to see was clear evidence, so it seemed, of tangible recombinant drugs, bottled, labeled, and apparently ready for sale and use in patient treatment. The prospectus text, restrained and restricted by a conservative SEC, told an unremittingly cautious story. The photos and diagrams told another: this pioneering company appeared on the verge of producing a revolutionary class of pharmaceuticals tailored specifically for human use. The media picked up on Genentech's seeming ability to carry through on this groundbreaking promise and, according to Middleton, developed the pending IPO "into a frenzy."[89]

In September Swanson, Perkins, Boyer, Middleton, and three representatives of the lead investment banks, Blyth Eastman Paine Webber and Hambrecht & Quist, made the rounds of American and European stockbrokers and pension and mutual funds. In a grueling series of investor meetings called "road shows," they pitched Genentech to the money managers in resolute efforts to sell blocks of stock. City after city, country

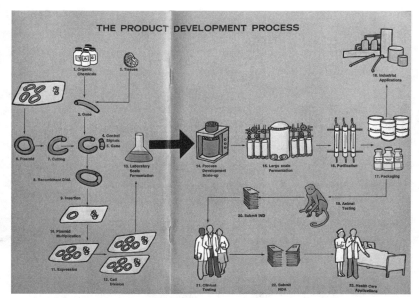

Fig. 17. Diagram of "The Product Development Process." (Initial public offering prospectus, Genentech, Inc., October 14, 1980.)

after country, the team trotted out an upbeat story of Genentech's prospects, finding an overwhelmingly enthusiastic response everywhere it went. Middleton recalled:

> People just listened and gaped. Herb [Boyer] got up and did his trick with the pop beads, showing how recombinant DNA works. . . . Basically, we had a little clear plastic box with pop beads in it—the baby toys that popped together. [The box] was supposed to represent a bacterium. He took out the beads and showed how you spliced genes together. A very simplistic little model. The fact that a UCSF professor was up there explaining it had everyone mesmerized. I gave the talk on the financial side, Bob gave the talk on the strategy, Herb gave the talk on the technology. It was pretty elementary. Every time we asked for questions, there weren't any. People didn't know what to ask. There were no experts, there were no analysts. Everybody was just amazed.[90]

The lengths to which the team went to explain the technology were laughable, but the pitches worked in conjuring up images of miracle drugs. A simple model and straightforward presentations allowed imaginations—and financial expectations—to soar. The *Economist*, in an article entitled "Frenetic Engineering," commented that the "mystique of genetic engineering" to create "medical wonders" had built an extraordinary market for the shares.[91] A Hambrecht & Quist banker commented, "We've never seen interest like this before."[92] Unfazed by Genentech's shaky valuation, lack of products, and multiple risks, brokers and investors snapped up all available shares, leaving many empty-handed and vociferously disappointed.

For Swanson, the road show's timing was far from perfect. He had married in September and, with the IPO delay, found his honeymoon unexpectedly intersecting with the road show. He described the improbable situation:

> Judy and I got married in Florida on September 2, a Thursday, I think. We had a weekend together in Paris, and then we were joined by seven men for the road show through Europe. It was Snow White and the Seven Dwarfs going through Europe. We did two cities a day. From Tuesday through Friday we did Paris, Geneva, Zurich, Edinburgh, Glasgow, London. We'd arrive [in a city] and we'd tell my wife, "We'll be back in two hours. . . . Have fun looking around."[93]

Fig. 18. Fred Middleton and Herb Boyer, on a break during the IPO road show, Zurich, Switzerland, September 1980. Compare this image of Boyer the entrepreneur with the image of Boyer the scientist on p. 14. (Photographer unknown; photograph courtesy of Fred A. Middleton.)

The unpropitious start to wedded life could have carried a dire warning, but a happy marriage and the birth of two beloved daughters ensued.

THE IPO

By October excitement over Genentech's imminent offering had reached fever pitch. Investors breathlessly awaited a chance to buy a piece of hot biotech action. On October 14, shortly after market opening, Genentech, under the NASDAQ stock symbol GENE, offered 1.1 million shares for sale. The last-minute increase in shares from 1 million indicated the strength of investor demand. A minute after the opening bell, the share price sky-rocketed from $35 to $80—the fastest first-day gain in Wall Street history. The price peaked at $89 within twenty minutes, rose and fell over the course of the day, and ended at $71 at market close. Genentech had raised over $36 million.[94] Based on the closing price, the company's value was an estimated $532 million.[95] It was, Perkins recalled, "the hottest stock offering in history to that time."[96] Even the staid *Wall Street Journal* called it "one of the most spectacular debuts in memory."[97] The explosive run-up and frenzied response flung the stock market back on its heels, the acceleration in share price prompting headlines worldwide. An officer at a major brokerage commented, "I have been with the firm 22 years. I have never seen anything like this."[98] But not only financiers were impressed. An article in *Science*, titled "Gene Splicing Company Wows Wall Street," reflected the science community's astonishment at investors' stampede to buy stock in a genetics-based company.[99]

Boyer and Swanson, holding 925,000 shares apiece, became instant multimillionaires, each reaping a one-day paper profit of nearly $70 million. Boyer, whose salary as a full professor was around $50,000, rushed out to purchase a Porsche Targa.[100] Swanson, "the first boy millionaire of biotech," as *Esquire* magazine dubbed him, hurried home to celebrate with his new and much wealthier bride.[101] The founders' initial $500 investments in Genentech had vaulted the sons of a railroad man and an airplane mechanic to an inconceivable peak of fame and fortune. For Kleiner & Perkins, Genentech's first investor, its 938,000 shares purchased at a reported average of $1.85 each were, in the baseball terms Perkins appreciated, a phenomenal home run.[102] Twenty-six-year-old, $10,000-a-year graduate student Richard Scheller dug out his stock certificates and found he was worth more than a million.[103] It was reason to celebrate with another reefer. Not everyone, however, made a killing. Ullrich was among the few who lost out. "I had this car that I bought for $250 which was breaking down all the time. So I decided I would buy a used VW Rabbit. So [before the IPO] I sold, I think, eight hundred shares for eight thousand dollars. . . . After we had gone public, the stock price went up and up and up. At some point, these eight hundred shares were worth more than a million dollars. And I bought a used Rabbit for that, a million-dollar Rabbit. Oh god!"[104]

The media had a field day covering the stunning public entrance of a company with the aura of revolutionary medical cures and premonitions of a new industry. Reporters and camera crews—CBS anchorman Walter Cronkite and the BBC among them—packed into the unprepossessing "world" headquarters of Genentech, Inc., doubtless unimpressed with the setting and only vaguely understanding the technology. What the media *did* understand was that everyone expected gene cloning to be "the cornerstone of a future billion-dollar industry."[105] The only black cloud was Stockholm's announcement, by strange coincidence on the very day of the IPO, of the award of a Nobel Prize in Chemistry to Paul Berg, Walter Gilbert, and Frederick Sanger for DNA manipulation and sequencing. The prize—and its glaring omissions—caught a nonplussed Stan Cohen by surprise and perhaps put a temporary damper on Boyer's elation.[106] Colleagues wondered how the Nobel committee came to ignore Cohen and Boyer's far more seminal invention.[107] Some speculate that Boyer's "going commercial" in founding and advising Genentech may have upset committee members and killed his chance, and consequently Cohen's, for receiving the prize. In November Cohen, Boyer, Berg, and Stanford colleague Dale Kaiser, received the Albert Lasker Basic Medical Research Award, for work laying the technical foundation for genetic engineering.[108] Prestigious though the Lasker was, it was not the Nobel Prize. With the singular power and reach of their method more evident with each passing year, Cohen and Boyer received the National Medal of Science in 1988 and 1990, respectively, and both received the National Medal of Technology in 1989.

For Genentech employees, the IPO was a startling revelation. To their amazement, the struggling start-up they had labored to keep afloat had gained an instant trove of cash and become the acknowledged front-runner in what could now be seen as an emerging industrial field. The offering's roaring success awakened them to the realization that the stock they had so casually accepted suddenly had real monetary value. It began to sink in that Genentech was not only a place to do frontier science but also where a scientist could make money, perhaps big money. DNA had acquired highly visible dollar signs in more ways than one. Dan Yansura recalled his reaction:

> My first thought on that day [of the IPO] was that Genentech went from being a research boutique to becoming a "real company" with publicly traded stock. Up till that point, money for supplies or for our paychecks was not a guaranteed thing. . . . Now all of a sudden we had a financial cushion to rest on a bit. . . . There was also an excitement and pride about being at the cutting edge of this new field. . . . Of course the last thoughts were about [my] own financial reap for the stock that I owned. This was more money that I ever thought of attaining at that point in my life. And just like that, there it was—kind of shocking.[109]

The end of the year brought another encouraging event. On December 2, 1980, the U.S. Patent Office issued patent 4,237,224 on a "Process for Producing Biologically Functional Molecular Chimeras." It was the now-famous pioneering patent on the Cohen-Boyer recombinant DNA procedure, the first major patent in biotechnology. The very fact that it issued (after six trying years of major political and legal obstacles) built confidence that the government would indeed grant broad patent protection on fundamental inventions at the heart of biotechnology. By December 15 seventy-two companies, including Genentech, had licensed the technology from Stanford, for a relatively low annual fee of $10,000.[110] Following within six months of the *Chakrabarty* decision, the patent's approval and its rapid licensing gave further reassurance that intellectual property in biotechnology could be legally protected and successfully licensed. More than that, the patent gave Cohen and Boyer's invention legitimacy and potency in the eyes of the law. Although problems regarding the viability and legal standing of patents in biotechnology were far from completely resolved, the Cohen-Boyer patent's wide scope, dominant position, and reasonable licensing fee were stabilizing and reassuring factors in the heady but uncertain business and legal environment of biotechnology in the early 1980s.[111]

Public demand to own shares in the expected biotech bonanza remained for a time sky-high and had a spill-over effect on other industries. Genentech's blazing good fortune exhibited to a rapt audience that a company without products and substantial revenues could nonetheless raise impressive amounts of public money. Executives of entrepreneurial, research-based companies took note of this significant departure from conventional business practice. A few abruptly changed their business plans. The *Economist* predicted—correctly, it turned out—that the lively action in Genentech stock would "whet appetites for other glamorous share offerings expected soon."[112] Appetites were indeed whetted. Investors were reported to be "positively salivating" over Apple Computer, which, following Genentech's lead, went public in December 1980.[113] By then Genentech's stock price had dipped below $45, nearer the offering price of $35. But the company had already made its mark as a trailblazer and inspiration for others.

Cetus cofounders Ron Cape and Pete Farley avidly took note. Farley had stated as recently as August 1980 that his company had no thought of going public until at least mid-decade.[114] In the wake of Genentech's precedent-breaking IPO, he and Cape had a sudden change of heart; they precipitously dropped the previous timetable and rushed out plans for a March 1981 IPO. Perkins noted the fertile seeds Genentech had sown: "[Genentech's IPO] established the idea that you could start a new biotechnology company, raise obscene amounts of money, hire good employees, sell stock to the public. Our competitors started doing all of that."[115] The 1980–81 period would see the creation of a fleet of entrepreneurial biology-based companies—Amgen, Chiron, Calgene, Molecular Genetics, Integrated Genetics, and firms of lesser note—all inspired by Genentech's example of a new organizational model for biological and pharmaceutical research. Before the IPO window closed in 1983, eleven biotech companies, in addition to Genentech and Cetus, had gone public.[116] Here was the gold rush Nicholas Wade predicted.

But not only institutions were transformed. Genentech's IPO transformed Herb Boyer, the small-town guy of blue-collar origins, into molecular biology's first industrial multimillionaire. For admiring scientists laboring at meager academic salaries in relative obscurity, he became a conspicuous inspiration for how their own research might be reoriented and their reputations enhanced. If unassuming Herb—just a guy from Pittsburgh, as a colleague observed—could found a successful company with all the rewards and renown that entailed, why couldn't they? The media reinforced Boyer's image of the astoundingly successful scientist-entrepreneur. In 1981 he was one of four runners-up to President Ronald Reagan in *Time* magazine's man of the year.[117] Two months later Boyer's portrait appeared on a *Time* cover, accompanied by the by-line, "Shaping Life in the Lab—The Boom in Genetic Engineering."[118] His beaming face and mop of unruly curls fading into a background of DNA helixes suggested his close and lucrative link with the genetic substance. Boyer had become an icon of biotechnology.

Not everyone was happy with the new commercialism, or with Boyer and colleagues themselves. To some, the flip side of the icon was mercenary scientists turning research funded by the public into private commercial and personal gain. The contention was but an early stage in an issue still debated in biotechnology: how best to balance basic and commercial interests in academic research?[119] But for the outside world of 1980, it was the dazzling vision of what commercial biotechnology might accomplish in medicine, energy, agriculture, and chemical production that riveted imaginations and stoked dreams. Academic scientists from now on would increasingly seize opportunities to put basic research to practical social benefit and in doing so gain possible riches and the satisfaction of seeing their research made useful to society.

A path had been broken for the growth of a far-flung, interactive network of relationships between academic biology and an expanding fleet of biotechnology companies.[120] Although university-industry associations were in no way new, either in the United States or abroad, what *was* new was the explosive growth and significance of such connections in biomedical research from the 1980s onward. In this, Genentech, in the persons of Boyer and Swanson and their young acolytes, had led the way. As early as 1983, an observer of an emerging biotechnology industry could claim with some accuracy: "Just about every leading molecular biologist in the United States has some form of industrial consultancy, financial investment in a new biotechnology or direct salaried involvement."[121] It was not a situation that every academic took lightly. The variety of perspectives on the new commercialism entering American life science was as varied as the individuals expressing them. One thing, however, was certain: for the first time in the history of their discipline, molecular biologists had a plethora of discoveries that might be commercialized and the incentives to do so. The age of the entrepreneurial biologist had arrived.[122]

Genentech's triumphal public debut was neither predictable, predetermined, nor inevitable, whether taken from the standpoint of tech-

MARCH 9, 1981

$1.50

TIME

Shaping Life in the Lab

Prince Charles Picks a Bride

The Boom In Genetic Engineering

Genentech's Herbert Boyer

Fig. 19. *Time* magazine cover of Boyer, March 9, 1981. (From TIME © March 9, 1981, TIME, Inc. All rights reserved. Used by permission and protected by the Copyright Laws of the United States. The printing, copying, redistribution, or retransmission of the Material without express written permission is prohibited.)

nology, politics, cultural precedents, social norms, or the variable factors of human motivation and performance. Important as recombinant DNA techniques were, Genentech's early evolution, social impact, and significance for a new industrial sector were emphatically *not* centered solely upon its technology.[123] On the contrary, as this history illustrates time and again, Genentech and the origins of biotechnology were far more than the successful industrial application of a novel technology. A concatenation of political, social, and economic factors and strategic scientific, financial, and business decisions molded, shaped, stymied, and encouraged Genentech's rise to the temporary pinnacle of its stock market debut. The company's future would be a roller-coaster ride of business ups and downs, vanguard medical achievements and disastrous decisions, aggressive patenting and incessant litigation. But as 1980 drew to a close, Genentech counted as a technological, corporate, and cultural experiment that against the odds had turned out well enough to serve as a bellwether and malleable template of a biotechnology industry about to emerge.

Epilogue

Genentech's sensational public offering conveyed more than the arrival on a spotlit stage of a transformational technology. The firm was the clear and acknowledged front-runner in commercial biotechnology. On significant fronts, Genentech had gotten there first and led the way—in financing and industrializing a basic-science procedure, organizing a new kind of business and legal structure around it, and winning scientific, corporate, and public investment and acclaim. Its founding generation had reaped the advantages of pioneering a virgin industrial field— free choice of research projects, first option on hiring appropriately trained scientists and managers, no intimidating wall of biotech patents to scale, and an identity and set of accomplishments setting it apart from competitors. In so doing, the company created a new and expansive social and cultural arena for economic activity in biomedical and pharmaceutical research.

Genentech exemplified a novel corporate form and milieu for business in biology, notably different from anything the pharmaceutical industry offered: the small, fleet, entrepreneurial firm in which innovative, close-to-basic research—often equal or arguably superior in quality to that of top academic labs—was the dominant focus and effort. Its culture, melding academic and corporate attributes and penetrating traditional institutional boundaries, bespoke biotechnology's uniquely close conceptual, material, and human interconnections with the research university. The firm's scientific successes helped to erase the stigma associated with industrial research and inspired a long line of academics to create, join, or consult for biotechnology companies.

Genentech's imprint on the shape of the biotechnology industry to come was broad and consequential. By 1980 it had blazed a trail that many young companies would follow: from venture capital–supported start-up, to contract-research organization, and, if fortunate, to public company with a seat on a stock exchange. Genentech provided organizational prototypes and cultural practices that entrepreneurs and scientists could use as starting points for loose emulation and adaptation. When cofounding Amgen in 1980, George Rathmann recalled that he turned to Genentech for inspiration for "how you get the job done," heeding its templates for raising money and forming corporate alliances.[1] Edward Penhoet, cofounder of Chiron in 1981, likewise acknowledged its status as exemplar and trailblazer:

> There's no question [Genentech] was a model for many of us. First of all, Genentech had already successfully traversed a lot of the ground that we would eventually cross ourselves. They had filed a lot of patent applications, so they had defined, in a sense, the intellectual property landscape [in biotechnology]. They had raised a lot of money, so they proved it could be done. They had recruited outstanding scientists.[2]

Underlying Genentech's achievements was the prime importance of its people. As Swanson profoundly appreciated, they were the firm's most valuable resource. "Most of our technology," he was wont to repeat, "walks out every night in tennis shoes."[3] And because research was the heart of the company, the basis upon which the early enterprise would sink or swim, it was the first generation of scientists who were singularly important to the firm's success and future viability. Again, it was Swanson who commented: "If the research goes well, we can handle the rest of the problems of the world."[4] The firm's freewheeling, go-for-broke culture—an electric distillation of individual high energy, creativity, competitiveness, and hubris—not only helped to keep scientists and managers coming to Genentech but also counted as a significant ingredient and treasured asset. Swanson's unflagging insistence on product focus and fiscal responsibility kept industrially inexperienced scientists supported and in line with business objectives and the firm headed toward products, patents, and profits. While Cetus and Biogen executives held a loose rein and flung talent, time, and money over a range of diverse projects, Swanson sailed close to the wind, keeping the early company steadily on track and advancing, financing round by financing round, project by project, corporate alliance by corporate alliance. Although business and governance would assume ever-greater roles in the public company, the fundamental centrality of science would endure as a distinguishing characteristic of Genentech and the biotechnology industry more generally.

By 1985 the biotechnology industry had attracted over $3 billion in capital, and industries as diverse as agriculture, fine chemicals, and pharmaceuticals had at least some genetic engineering capacity.[5] Those companies premised on pharmaceutical manufacture found Genentech's 1970s production of one recombinant protein after another difficult or impossible to emulate and sustain (as did Genentech itself). With the advantage of a front-runner position in a virgin field, its scientists had plucked the low-hanging fruit—replacement proteins for existing drugs—and left to its competitors, and to itself, the difficult science of making novel and complicated pharmaceuticals, such as Amgen's Epogen (a stimulant of blood cell formation) and Genentech's tissue plasminogen activator (tPA, a dissolver of blood clots). The young industry found that marketable products were surprisingly difficult to realize, financing often fickle and short-term, regulatory agencies more demanding than expected, litigation commonplace and costly, and public policy and the economic climate fluctuating and undependable. Genentech itself foundered in the mid-1980s, undergoing profound financial problems and difficult transitions at the executive level. Swanson was forced out, Kirk Raab stepped up, and its overhyped tPA failed to meet market projections. In 1990 Hoffmann–La Roche rescued the company in a 60 percent buyout, with a future option to purchase the entirety.

The troubles accompanying the rise of a new industry were not only internal. One of the most immediate was the relationship of the budding biotechnology industry and the American research university. The tight bonds that Genentech had formed with the University of California in the 1970s and that later start-ups would duplicate with other universities sent both expectant and anxious tremors down the corridors of academe. The flight of university biologists into biotechnology companies, the surge in faculty consulting, and acceleration in university patenting and licensing that began in earnest in the 1980s (and continue full force in the twenty-first century) generated long-winded debate and contention within and outside academia as the traditional walls between the two spheres grew increasingly porous. The ivory tower of academic biomedicine was being scaled, certainly not for the first time, but in the 1980s with an energy and pervasiveness that was entirely new to the discipline.

American research universities, particularly those with strong programs in molecular biology, were for a time in turmoil as they struggled to develop policies to balance academic traditions of scientific openness and exchange with the privatization and proprietary claims accompanying the mounting industrialization of the American research university. Although university-industry associations were in no way new, either in the United States or abroad, what *was* new was the explosive growth of such connections in U.S. biomedical research of the 1980s. *The Nation* observed, just one voice among the many: "Academic biologists, who have traditionally defended the purity of their research, can no longer claim to be white-coated keepers of objective truth. Like the institutions they work for, they have clear economic interests to protect."[6] As the Boyers, Gilberts, and Weissmanns—to be followed in the 1980s by an interminable string of academic entrepreneurs, often in lockstep with fervid venture capitalists—built bridges between molecular science and the industrial world, they experienced the exhilaration of making their research practical and sometimes highly profitable but also the professional and personal trials of plowing commercial ground for which American biomedical research institutions had inadequate guidelines. As such ties became commonplace and widely accepted in American academia, industrial, economic, and proprietary interests would breach the fortress of academic biology to an extent and intensity previously unknown, carrying numerous practical social benefits, worrying ethical concerns, and profound cultural and attitudinal changes that society continues to both welcome and debate.

In the decades ahead, Genentech would misstep, suffer failed projects, over-promote products, endure bad press, see its libertine culture tamed, and undergo the challenges of corporate expansion, fierce competition, and near-constant litigation. Yet the fundamental principles upon which Boyer and Swanson created the company—the primacy of research, a focus on pharmaceutical manufacture for human application, encouragement of individual creativity, and a university-like culture—remain the foundation upon which the twenty-first-century company operates. Under the direction of Arthur Levinson, third chairman and chief executive, Genentech could look back on thirty-some years of productivity and a current line of much-used therapies for cancer and other difficult-to-treat diseases. In 2009 Hoffmann–La Roche acquired the remaining 40 percent of the company for $47 billion.

EPILOGUE

Notes

CHAPTER 1

1   Boyer oral history, 1994, 41.
2   Boyer, transcript of talk, 1992. "Winding Your Way through DNA" symposium, University of California, San Francisco, September 25–26, 1992.
3   Cohen oral history, 2009, 2.
4   Boyer interview, 2009, 3.
5   Boyer interview, 1975, 4.
6   Ibid.
7   J. Michael Bishop, introduction to Boyer oral history, 1994, x.
8   Cohen oral history, 2009, 8.
9   For a history of the laboratory use of restriction enzymes, see Roberts 2005.
10 Boyer interview, 1975, 9.
11 Boyer interview, 2000, 20.
12 Much early work in molecular biology focused on bacterial viruses (bacteriophage). See Cairns, Stent, and Watson 1966.
13 For a history of bacterial drug resistance, see Creager 2007.
14 For a history of the use of plasmids in research, see Cohen 1993 and his oral history, 2009, 25–30.
15 Cohen oral history, 2009, 18.
16 Kornberg oral history, 1997, 22–25.
17 Berg oral history, 1997, 113.
18 Cohen, e-mail message to author, June 28, 2010. For firsthand accounts of the science and culture of Stanford biochemistry in the 1970s and 1980s, see the Kornberg oral history, 1997, and Berg oral history, 1997. For Cohen's views on his relationship with Stanford biochemistry, see his oral history, 2009, 19–20.
19 Cohen, e-mail message to author, June 28, 2010.

20 Jackson et al. 1972.

21 Berg oral history, 1997, 70.

22 Boly 1982.

23 For a history of social accountability in genetics, see Weiner 1999.

24 Boyer interview, 1975, 19–21.

25 Boyer interview, 2000, 53.

26 For Rutter's strategy for revitalizing the UCSF biochemistry department, see his oral history, 1992, 20–43.

27 William J. Rutter, Annual Report, 1970, Department of Biochemistry and Biophysics, UCSF. For molecular biology's orientation to the study of higher organisms, see Morange 1997.

28 Boyer oral history, 1994, 25.

29 For Cohen's work on drug interactions, see his oral history, 2009, 22–23.

30 Ibid., 34–36; Cohen et al. 1972.

31 Mertz and Davis 1972; Sgaramella 1972.

32 Hedgpeth et al. 1972.

33 For a history of this work, see Berg oral history, 1997, and Yi 2008.

34 Cohen to Boyer, October 24, 1972, Cohen's office correspondence, Stanford University.

35 Cohen oral history, 2009, 43–48.

36 Cohen and Boyer concur that the walk to the deli occurred after a day of conference presentations. Stanley Falkow, a participant in the walk, believes that it happened on the evening before the conference began. Boyer, e-mail message to author, January 18, 2010; Falkow, e-mail message to author, August 1, 2009; Falkow 2001.

37 Boyer, videotaped interview produced for an educational series, "Winding Your Way through DNA," undated transcript [1994], tape 9, p. 3.

38 Cohen oral history, 2009, 48.

39 Quoted in Boly 1982.

40 Sharp et al. 1973.

41 Boyer oral history, 1994, 34–35.

42 Cohen oral history, 2009, 49.

43 For accounts of this and the two subsequent experiments, see ibid., 49–52, and Cohen 1975.

44 The Boyer quotation comes from "Biotech's Beginnings: Genenlab Notebook," a humorous facsimile of a laboratory notebook produced by Genentech on its twentieth anniversary.

45 Cohen oral history, 2009, 51.

46 Boyer interview, 1975, 15.

47 Cohen et al. 1973.

48 The relevant sentence is: "The general procedure described here is potentially useful for insertion of specific sequences from prokaryotic [microbial] or eukaryotic [animal] chromosomes or extrachromosomal DNA into independently replicating bacterial plasmids."

49 Cohen oral history, 2009, 55. Boyer remarked: "I don't remember that ar-

rangement with Stan. However, Stanley was reasonably secretive about what he was doing." Boyer oral history, 1994, 102.

50 Falkow interview, 1976, 73–74.

51 Betlach oral history, 1994, 14.

52 Quoted in Lear 1978, 70.

53 The sentence reads: "The replication and expression of genes in *E. coli* that have been derived from a totally unrelated species . . . now suggest that interspecies genetic recombination may be generally obtainable. Thus, it may be practical to introduce into *E. coli* genes specifying metabolic and synthetic functions (e.g., photosynthesis, antibiotic production) indigenous to other biological classes." Chang and Cohen 1974, 1033.

54 Boyer oral history, 1994, 107–10. Some thirty years later, Berg and Boyer hotly debated whether Boyer or Morrow suggested the frog experiment. Paul Berg contended that it was Morrow, who, having found that EcoR1 cut the frog DNA once, approached Boyer about "inserting the [frog] rDNA samples into Cohen's plasmid." Berg, e-mail message to author, October 27, 2008. Boyer, after reading Berg's e-mail assertion, responded: "Actually at that conference I talked to John Morrow about cloning eukaryotic [animal] DNA and the difficulties that might entail. Contrary to the recent article by Berg and Mertz, he [Morrow] did not approach me but rather shared the information that EcoR1 cut *X. laevis* ribosomal DNA that he had received from [his adviser] Don Brown. He offered to provide some to me and I suggested that he should be a coauthor if the experiment should be successful." Boyer, e-mail message to author, October 5, 2009. Berg repeated his claim in Berg and Mertz 2010.

55 Cohen, e-mail message to author, October 5, 2009.

56 Quoted in Kathryn Christensen, "Gene Splicers Develop a Product: New Breed of Scientist-Tycoons," *Wall Street Journal*, November 24, 1980.

57 Boyer, "Winding Your Way through DNA," tape 8, p. 8.

58 Boyer oral history, 1994, 46–47.

59 Morrow et al. 1974.

60 News release, Stanford University News Service, May 20, 1974, patent correspondence S74-43, 1974–80, Office of Technology Licensing, Stanford University.

61 See, for example, Hotchkiss 1965.

62 "Getting Bacteria to Manufacture Genes," *San Francisco Chronicle*, May 21, 1974, 6.

63 "The Gene Transplanters," *Newsweek*, June 17, 1974, 54.

64 Treatment here of the Cohen-Boyer patent history is based on the more comprehensive history in Hughes 2001.

65 Victor K. McElheny. "Animal Gene Shifted to Bacteria; Aid Seen to Medicine and Farm," *New York Times*, May 20, 1974.

66 Cohen 1982.

67 Paul Berg and Janet Mertz 2010 remarked: "The sociology among most U.S. life scientists prior to the 1970s was to eschew patents, believing that they

would restrict the free flow of information and reagents and impede the pace of discovery" (15).

68 For a history of controversy over university patenting in biomedicine, see Weiner 1986.

69 Boyer oral history, 1994, 116.

70 Donald R. Helinski to Josephine Opalka, July 1, 1975, patent file 74-134-1, Cohen, S. et al., Office of Technology Transfer, University of California. Helinski went on to describe the patent application as "ill-conceived" because it disregarded "the contributions of other scientists and is very basic in its concepts and applications."

71 For the history of the recombinant DNA controversy and the debate over research regulation, see Krimsky 1985. Watson and Tooze's *The DNA Story* (1981) consists of a collection of contemporary documents with commentary on the recombinant DNA controversy and gene cloning.

72 For Berg's role in the political controversy, see his interview, 1978, and oral history, 1997.

73 Cohen oral history, 2009, 69.

74 For a comprehensive history of recombinant DNA research regulation in the United States and the United Kingdom, see Wright 1994.

75 Cohen oral history, 2009, 89–92.

76 Boyer oral history, 1994, 46.

77 Berg interview, 1978, 71.

78 Cohen, telephone call to author, November 17, 1994.

79 For discussion of the growing public demand for greater practical value from scientific research, see Vettel 2006, chaps. 5–7.

80 For Cape's biography and account of Cetus's foundation and development, see his oral history, 2003.

81 Quoted in Lewin 1978b, 19.

82 For Stanford's long-standing industrial connections, see Lowen 1997.

83 Lederberg 1975, 33.

84 "Policy on Consulting by Members of the Academic Council: Principles and General Standards," March 11, 1977, Arthur Kornberg Papers, SC359, box 5, folder: "1977," Green Library, Stanford.

85 Cetus archive, box Co120181581, folder: "Consultants: Stanley Cohen," Chiron Corporation (Chiron acquired Cetus in 1990). Author's access to Cetus and Chiron documents at Chiron Corporation courtesy of William Green, former Chiron general counsel.

86 Mary Betlach, a technician in the Boyer laboratory, remarked, "I was doing a *lot* of cloning of DNA from all kinds of organisms. It was like the zoo." Betlach oral history, 1994, 16.

87 Boyer interview, 2000, 9.

88 Bill Carpenter to File S74-43, memo re: Dr. Herbert Boyer, September 18, 1974, S74-43, patent correspondence 1974–1980, Office of Technology Licensing, Stanford.

89 Boyer oral history, 1994, 71.

90 Boyer interview, 2000, 9.

91 Boyer interview, 1975, 35.

92 Ken Imatani to Reimers, memo, "Discussion with Dr. Boyer for Future
    Development Work for Recombinant DNA Process," August 6, 1975,
    S74-43, patent correspondence 1974–1979, Office of Technology Licensing,
    Stanford.

CHAPTER 2

1   Boyer oral history, 1994, 72.
2   Swanson oral history, 1996/1997, 3.
3   Kenneth P. Morse, introduction to ibid., x–xiv.
4   Swanson oral history, 1996/1997, 2.
5   Ibid., 5.
6   Ibid.
7   For Silicon Valley's history, culture, and major figures, see Lécuyer 2006.
8   In the late 1970s, Kleiner & Perkins became Kleiner Perkins Caufield &
    Byers, its current name. For the early history of the firm, see Perkins 2007
    and his two oral histories, 2001 and 2009.
9   Perkins oral history, 2001, 3.
10 Years later Perkins recalled that Kleiner & Perkins's investment in Cetus
    was "probably half a million. Which was a big deal in those days." Ibid., 4.
11 Ibid. Also see Perkins 2007, 119–25.
12 For Glaser's opinion of Cetus's missed opportunity in recombinant DNA
    technology, see his oral history, 2003–4, 107–8.
13 Perkins oral history, 2009, 45.
14 Perkins oral history, 2001, 3.
15 For information on Cetus's Recombinant Molecular Research Division,
    see Gelfand interview, 1978, 34–39, 55–58, passim; and Rabinow 1996,
    41–44. Cohen tried without success to interest Cetus in trying "to express
    genes for human hormones in bacteria." Cohen oral history, 1995, 111.
16 Cape interview, 1978, 30.
17 "Recombinant DNA Research Guides Worry Drug, Chemical Industry,"
    Blue Sheet 19, no. 23 (June 1976). The pharmaceutical companies repre-
    sented were Eli Lilly, Roche Institute of Molecular Biology, Merck, Up-
    john, Wyeth, Abbott, Pfizer, Smith Kline & French, Ciba, and Burroughs
    Wellcome. Cetus also attended.
18 Perkins oral history, 2001, 3.
19 Swanson oral history, 1996/1997, 21.
20 Ibid., 10.
21 Ibid., 13.
22 Quoted in Rothenberg 1984, 370.
23 Robert Swanson, "Stanford Speech," draft of speech accepting an Entre-
    preneurial Company of the Year Award to Genentech, Stanford Business
    School, 1983, Swanson's office correspondence, K & E Management, San
    Mateo, California.
24 Swanson oral history, 1996/1997, 16.

25 Swanson, "Stanford Speech."

26 Ibid.

27 Boyer oral history, 1994, 72.

28 Quoted in Rothenberg 1984, 371.

29 Boyer oral history, 1994, 95–96.

30 Ibid., 71.

31 Ibid., 72.

32 Ibid.

33 Swanson, "Stanford Speech."

34 See, for example, Marx 1976.

35 "Genentech, Inc., Business Plan, December, 1976," Chief Financial Officer files, Genentech, Inc.

36 Swanson unsuccessfully pitched venture investment in Genentech to Charles Crocker. "Outline for Discussion, Crocker Capital, March 12, 1976," Chief Financial Officer files, Genentech, Inc.

37 Quoted in Sylvester and Klotz 1983, 87.

38 "Outline for Discussion, Kleiner & Perkins, April 1, 1976," Chief Financial Officer files, Genentech, Inc.

39 One estimate placed the cost in 1975 of moving a candidate drug through the development and approval process at $138 million. Jones 2005, 2:545–46.

40 Perkins oral history, 2001, 4–5.

41 For history of the first generation of venture capitalists in California, see the oral history series by the author, including interviews with Perkins, at http://bancroft.berkeley.edu/ROHO/projects/vc/.

42 Eugene Kleiner to Nathaniel I. Weiner, May 7, 1976, box 342652, folder: "Genentech," Chiron Corporation.

43 Quoted in Sylvester and Klotz 1983, 87.

44 Swanson, "Stanford Speech."

45 "Genentech, Inc., Business Plan, December, 1976."

46 "Genentech, Inc., Financial Statements, Period from Inception (April 7, 1976) to December 31, 1976, with Report of Certified Public Accountants," Chief Financial Officer files, Genentech, Inc.

47 Perkins oral history, 2001, 24.

48 Ibid. Perkins resigned as board chairman in 1990 and as a director in 1995.

49 The description of the microbiology department comes from the Rutter oral history, 1992, 18.

50 "Genentech [Business Plan], Meeting, Stanford Office of Technology Licensing," April 19, 1976, Chief Financial Officer files, Genentech, Inc.

51 Boyer and Swanson to "Gentlemen, Stanford University, University of California," April 19, 1976, 74-134-1, folder: "Cohen-Boyer, Exploitation," Office of Technology Transfer, University of California.

52 John K. Poitras of Stanford's Office of Technology Licensing quoted Farley in a memo to the Cohen-Boyer patent file. "JKP to S74-43," Correspondence 1974–1980, Office of Technology Licensing, Stanford University.

53 "Genentech [Business Plan], Meeting, Stanford Office of Technology Licensing."

54 Reimers to Robert Augsburger, July 19, 1976, S74-43, Correspondence 1974–1980, Office of Technology Licensing, Stanford University.

55 Swanson oral history, 1996/1997, 34–35.

56 Perkins oral history, 2001, 8. In August 1981 Stanford announced the availability of licenses on the Cohen-Boyer invention. Genentech was among the seventy-two companies to obtain licenses by the end of that year. Hughes 2001.

57 Perkins oral history, 2001, 24.

58 Kiley oral history, 2000/2001, 15.

59 "Financial Statements, Genentech, Inc., Period from inception (April 7, 1976) to December 31, 1976."

60 "Genentech, Inc, Business Plan, December 1976."

61 Ibid.

62 Ibid.

CHAPTER 3

1 Handler quoted in untitled announcement of somatostatin research, December 2, 1977, UCSF News Services/Publications, University of California, San Francisco.

2 Herbert Boyer, "The Business," symposium paper, "The Emergence of Biotechnology: DNA to Genentech," Chemical Heritage Foundation, Philadelphia, June 13, 1997, 80–92.

3 Greene et al. 1975; Howard M. Goodman, associate professor of biochemistry, and Herbert W. Boyer, associate professor of microbiology, Annual Reports, 1974, Department of Biochemistry and Biophysics, UCSF. Because Boyer and Goodman had a cooperative research arrangement, Boyer's research was included in the Department of Biochemistry's annual reports, even though he did not officially join the department until December 1975.

4 Boyer, e-mail message to author, January 11, 2010. Riggs's memory of the episode differs. He recalls that he telephoned Boyer, who agreed to provide his DNA linkers. Riggs, e-mail message to author, January 8, 2010.

5 Heyneker et al. 1976.

6 Heyneker oral history, 2002, 57.

7 Ibid.

8 "Synthetic DNA Put to Work in Living Cells," October 28, 1976, AR86-7, carton 2, News Services: Records 1976–1986, UCSF Library and Center of Knowledge Management.

9 Boyer, e-mail message to author, January 8, 2010.

10 Boyer oral history, 1994, 73.

11 Itakura oral history, 2005, 18–19.

12 Riggs oral history, 2005, 34.

13 Sidney Brenner, quoted in Wright 1986, 323–24.

14 Riggs oral history, 2005, 33.

15 Keiichi Itakura, "Human Peptide Hormone Production in *E. coli*," February 26, 1976, grant application, Department of Health, Education, and Welfare. Copy courtesy of Arthur Riggs.

16 Riggs oral history, 2005, 30.

17 Michael I. Goldberg to Itakura, November 3, 1976. Copy courtesy of Drs. Itakura and Riggs.

18 Swanson oral history, 1996/1997, 26.

19 Riggs oral history, 2005, 46.

20 Swanson oral history, 1996/1997, 22.

21 Sponsored Research Agreement, August 1, 1976, 77-064-1, folder: "Goodman et al., Rutter et al., Deposit of Microorganisms," Office of Technology Transfer, University of California, Berkeley; Michael Bader, "Scientist Splices Private Company, UCSF," *Synapse* (UCSF campus newspaper) 23, no. 9 (November 9, 1978).

22 Kiley oral history, 2000/2001, 9.

23 Contract appended to *City of Hope National Medical Center v. Genentech, Inc.*, Cal. App. 2d, Div. 2, B161549 (2003).

24 Kiley oral history, 2000/2001, 17.

25 "Sponsored Research Agreement by and between Genentech, Inc., and California Institute of Technology," October 1, 1976, Chief Financial Officer files, Genentech, Inc.

26 Scheller oral history, 2001/2002, 55.

27 *City of Hope v. Genentech*, 20 Cal. Rptr. 3d 234 (2004); Andrew Pollock. "Medical Center Seeks More Payments from Genentech," *New York Times*, August 29, 2001. For a summary of the case, see Rimmer 2009.

28 "Appellate Court Upholds Award to City of Hope," *Biotechnology Law Report* 24, no. 1 (February 2005): 51–54.

29 Bob Egelko and Bernadette Tansey, "Court Reduces Genentech Damages in Royalty Suit," *San Francisco Chronicle*, April 25, 2008.

30 Bernadette Tansey, "Genentech Loses $300 Million Royalty Battle: Punitive Damages Could Double Jury's Award to Medical Center," *San Francisco Chronicle*, June 11, 2002.

31 Swanson oral history, 1996/1997, 27. In addition to Kleiner & Perkins, the Mayfield Fund, International Nickel, Innoven Capital, Sofinnova, and Hambrecht & Quist invested at $2.89 per share. "Genentech, Inc., 1979 Corporate Plan," Chief Financial Officer files, Genentech, Inc.

32 Itakura oral history, 2005, 21–22.

33 "Guidelines for Research Involving Recombinant DNA Molecules," *Federal Register* 41 (July 7, 1976): 27911–43.

34 "Expression of Synthetic DNA In Vivo," Memorandum of Understanding and Agreement, February 9, 1977, AR86-7, carton 2, folder 76, UCSF Library and Center for Knowledge Management.

35 Rachmiel Levine, Executive Director, City of Hope, to Crea, May 16, 1977. Copy courtesy of Crea.

36 For Crea's arrival and research at City of Hope, see his oral history, 2002, 19–30.
37 Heyneker oral history, 2002, 44.
38 Riggs oral history, 2005, 63.
39 Itakura oral history, 2005, 32–33.
40 For details of the somatostatin research, see Heyneker oral history, 2002, 44–50, and the published paper Itakura et al. 1977.
41 Swanson oral history, 1996/1997, 36.
42 Riggs oral history, 2005, 41.
43 Swanson oral history, 1996/1997, 37.
44 The sentence reads, in a stretch of the truth: "The experiment was deliberately designed to have the cells produce not free somatostatin but rather a precursor, which would be expected to be relatively inactive." Itakura et al. 1977, 1063.
45 Itakura oral history, 2005, 35.
46 Quoted in Charles Petit, "A 'Triumph' in Genetic Engineering," *San Francisco Chronicle*, December 3, 1977.
47 Riggs oral history, 2005, 42–43.
48 Heyneker oral history, 2002, 62.
49 Swanson to shareholders, April 6, 1978, Chief Financial Officer files, Genentech, Inc.
50 Cape oral history, 2003, 32.
51 Kiley oral history, 2000/2001, 90. For the central role of patents in biotechnology, see Dutfield 2003, especially 152–54.
52 Itakura et al. 1977.
53 I thank Henry Bourne for this point.
54 Kiley oral history, 2000/2001, 37.
55 Riggs oral history, 2005, 49.
56 Genentech eventually obtained eleven U.S. and more than a hundred foreign patents on the technology—the important Riggs-Itakura patents, which are legal mainstays of the biotechnology industry.
57 Statement by Donald S. Fredrickson, MD, Director, National Institutes of Health on Recombinant DNA, before the Subcommittee on Science, Technology, and Space, Committee on Commerce, Science, and Transportation, United States Senate, November 8, 1977.
58 U.S. Senate, *Report on Regulation of Recombinant DNA Research, November 2, 8, and 10, 1977*, 95th Congress, 1st session, 1977, serial no. 95-52. For a synopsis of the hearing from the perspective of the NIH director, see Fredrickson 2001, 170–76.
59 "Testimony by Paul Berg," Subcommittee on Science, Technology and Space, November 2, 1977, Paul Berg Papers, SC 358, box 1, folder: "Senate testimony," Green Library, Stanford University.
60 Handler and Berg quoted in untitled announcement of the somatostatin research, for release December 2, 1977, UCSF News Services/Publications, University of California, San Francisco.

61 Victor Cohn, "An Artificial Gene Makes Copy of Brain Hormone," *Washington Post*, November 3, 1977.

62 "Human Gene in *E. coli*: It Works!" *Chemical and Engineering News*, November 7, 1977, 4.

63 See, for example, Victor K. McEleny, "Coast Concern Plans Bacteria Use for Brain Hormone and Insulin," *New York Times*, December 2, 1977, D1.

64 David Perlman, "Scientific Announcements," *Science* 198 (November 25, 1977): 782.

65 Riggs oral history, 2005, 47.

66 Quoted in Petit, "A 'Triumph' in Genetic Engineering."

67 Untitled announcement of the somatostatin research, for release December 2, 1977, UCSF News Services/Publications, University of California, San Francisco, and City of Hope publicity.

68 Quoted in Petit, "A 'Triumph' in Genetic Engineering."

69 Quoted in ibid.

70 See, for example, Bok 2003.

71 Bader, "Scientist Splices Private Company, UCSF."

72 Yamamoto 1982, 195–201.

73 Rutter oral history, 1992, 101–7.

74 For the foundation of Chiron, see Penhoet oral history, 1998.

75 Charles Petit, "The Bold Entrepreneurs of Gene Engineering," *San Francisco Chronicle*, December 2, 1977.

76 For the history of DNAX, see Kornberg 1995, chap. 2; Kornberg oral history, 1997; and Berg oral history, 1997.

77 A journalist remarked in 2007 on industry contracts with American academe: "The amount of activity has increased, and so has the complexity of the deals, but there's nothing new about professors selling their expertise off campus or performing industry-financed research in their university laboratories. They have been doing it at least since the Morrill Act of 1863 and follow-on legislation established the nationwide system of land-grant colleges, each with a mandate to promote 'agriculture and the mechanical arts.'" Greenberg 2007, 83.

78 Bliss 1982, 131–39; Swann 1988, chap. 5.

79 Chapter 4 treats the UC-Lilly contract on insulin with the Rutter and Goodman labs.

80 Sponsored Research Agreement, August 1, 1976, 77-064-1, folder: "Goodman et al., Rutter et al., Deposit of Microorganisms," Office of Technology Transfer, University of California, Berkeley.

81 Sheldon Wolff, Chairman, Committee on Rules and Jurisdiction, to Laurel E. Glass, Chairperson, Academic Senate, August 6, 1979, cover letter for "Report of Committee on Rules and Jurisdiction," submitted by Wolff, AR86-7, Genentech, carton 2, folder 94, UCSF Library and Center for Knowledge Management.

82 For a critique of such phenomena in the 1970s and 1980s, see Kenney 1986.

83 J. Michael Bishop, introduction to Boyer oral history, 1994, vii–ix.

84 Petit, "The Bold Entrepreneurs of Gene Engineering."

85 Boyer oral history, 1994, 98.

86 "A Commercial Debut for DNA Technology," *Business Week*, December 12, 1977, 128, 132.

87 Jeremy Rifkin, "DNA: Have the Corporations Already Grabbed Control of New Life Forms?" *Mother Jones*, February–March 1977, 23–26, 39.

CHAPTER 4

1 "Human Insulin: Seizing the Golden Plasmid," *Science News*, September 16, 1978, 195.

2 Boyer oral history, 1994, 88.

3 Middleton oral history, 2001, 17.

4 Johnson 1983.

5 Johnson oral history, 2004, 26.

6 Rutter oral history, 1992, 111; Rutter, e-mail message to author, February 5, 2010.

7 As mentioned below, Gilbert and colleagues founded Biogen in 1978.

8 Middleton to Swanson, April 6, 1979, Chief Financial Officer files, Genentech, Inc.

9 Perkins oral history, 2001, 24. After a few years in Berkeley, Cetus relocated to neighboring Emeryville, a city with an industry-friendly history.

10 Swanson to shareholders, April 6, 1978, Chief Financial Officer files, Genentech, Inc.

11 Untitled UCSF news release, for release May 23, 1977, 77-064-1, UC-1, Rutter, W., et al., Office of Technology Transfer, University of California.

12 Using the enzyme reverse transcriptase, scientists make a DNA copy from messenger RNA, the unstable transcript of genes. The backward flow of genetic information from RNA to DNA constitutes a reversal of molecular biology's central dogma: DNA makes RNA makes protein.

13 "One for the Gene Engineers," *Time*, June 6, 1977, 68.

14 Rutter oral history, 1992, 185–86; Josephine Opalka, UC patent administrator to A. R. Whale, Esq., Eli Lilly and Company, July 7, 1977, CU-6, UC (System), box 52, folder: "Science & Technology: rDNA Research," Office of Technology Transfer, University of California.

15 Chapter 5 discusses Seeburg's growth hormone research.

16 Ullrich oral history, 1994/2003, 21.

17 Robert A. Swanson to Axel Ullrich, April 1, 1977, Chief Financial Officer files, Genentech, Inc. The author has not located contracts for Seeburg and Shine.

18 Cohen oral history, 1995, 155.

19 Swanson oral history, 1996/1997, 54–55; Rutter oral history, 1992, 175.

20 Swanson oral history, 1996/1997, 25–26.

21 Swanson to Howard M. Goodman, April 1, 1977, Hormone Research Institute documents, UCSF. The author has not found a contract letter for Rutter.

22 Swanson to Axel Ullrich, April 1, 1977, Chief Financial Officer files, Genentech, Inc. The author has not found contract letters for Seeburg and Shine.

23 Rutter oral history, 1992, 175–76; Swanson oral history, 1996/1997, 25; Ullrich oral history, 1994/2003, 21–22.

24 *Regents of the University of California v. Eli Lilly & Co.*, 119 F. 3d 1559 (1997), Goodman cross-examination, August 29, 1995, 1235–36.

25 For firsthand accounts of the moratorium and its effects, see the Recombinant DNA Controversy Oral History Collection, MIT Libraries, Cambridge, MA. Also see, Krimsky 1985.

26 Walter Gilbert's laboratory, Biological Laboratories, Harvard University, *The Midnight Hustler* (newsletter) 1, no. 2 (May 1977). A subsequent note of June 1, 1977, informed Boyer lab tech Mary Betlach that she was in arrears for her newsletter subscription. The Gilbert lab's "Circulation Manager" (Forrest Fuller) asked her to remit "$4,000,000.00 or one insulin producing clone immediately." Copies courtesy of Mary Betlach.

27 Goeddel oral history, 2001/2002, 11.

28 Kleid oral history, 2001/2002, 25.

29 Kleid, telephone conversation with author, October 2, 2007.

30 Since the early twentieth century, major American industrial research laboratories had required their scientists to assign research rights to the company. See Wise 1980.

31 For a lively account of the contest for insulin at Harvard, UCSF, and Genentech, see Hall 1987.

32 Tom Kiley, letter to author, February 1, 2004.

33 Riggs oral history, 2005, 56.

34 Riggs 1981.

35 Crea oral history, 2002, 27.

36 Others working on human insulin at this time included Donald Steiner at the University of Chicago, Saran Narang in Canada, and Ray Wu at Cornell. "Bacterial Insulin Production Nears Reality," *Chemistry & Engineering News*, June 19, 1978, 4–5.

37 Kleid oral history, 2001/2002, 39.

38 Itakura oral history, 2005, 86.

39 Heyneker oral history, 2002, 69.

40 Goeddel oral history, 2001/2002, 26.

41 Heyneker oral history, 2002, 73–74.

42 "Bacterial Insulin Production Nears Reality," 4–5.

43 Robert Cooke, "Clone Business: It's Growing Fast, It's Growing Fast," *Boston Globe*, June 25, 1978. For the foundation of Biogen, see Weissmann 1981.

44 Yansura oral history, 2001–2, 25.

45 Quoted in Rothenberg 1984, 134.

46 Lécuyer 2006, 265.

47 Yansura oral history, 2001–2, 22.

48 Kleid oral history, 2001/2002, 44.
49 Yansura oral history, 2001–2, 28.
50 Ibid., 59.
51 Riggs oral history, 2005, 59–60.
52 Crea et al. 1978; Goeddel et al. 1979.
53 Kleid, e-mail message to author, March 26, 2007.
54 Ullrich oral history, 1994/2003, 29–30.
55 Richard Saltus, "Genetic Research—'I wouldn't say we're ahead in the race,'" *San Francisco Chronicle*, October 29, 1978, 1. The title quotes Bill Rutter.
56 Kiley oral history, 2000/2001, 35–36.
57 Denise Gellene, "Cancer Center, Drug Firm Fight over Biotech Riches," *Los Angeles Times*, August 28, 2001.
58 *Genentech, Inc. v. Eli Lilly and the Regents of the University of California* (1990) 90-1676C.
59 Boyer oral history, 1994, 79–80.
60 Swanson oral history, 1996/1997, 122.
61 Kiley oral history, 2000/2001, 64–65.
62 Johnson oral history, 2004, 39.
63 Kleid oral history, 2001/2002, 101.
64 Yansura oral history, 2001–2, 69.
65 Perkins oral history, 2001, 22.
66 Johnson oral history, 2004, 33–34.
67 Middleton oral history, 2001, 16–17.
68 Robert A. Swanson, "Genentech—A Commercial Record of Achievement in Molecular Biology," unpublished presentation to E. F. Hutton, New York, September 17, 1979, Swanson correspondence, K&E Management, San Mateo, CA.
69 For a study of such early interactions, see Kenney 1986.
70 A Harvard Business School professor contends: "The Genentech-Lilly agreement . . . created a template that influenced the evolution of the biopharmaceutical industry for the next thirty years." Pisano 2006, 86.
71 Goeddel oral history, 2001/2002, 51.
72 Ibid.
73 Hall 1987, 275.
74 Andreopolos 1980, 743.
75 The enormous family of Riggs-Itakura domestic and foreign patents issuing in the 1980s and '90s covers basic processes for producing recombinant proteins. They are fundamental to the biotechnology industry.
76 Boyer oral history, 1994, 87.
77 For a discussion of similar issues in early twenty-first-century biotechnology, see Kleinman and Vallas 2005.
78 Kiley oral history, 2000/2001, 19.
79 Reich 1985, 110, 118–20, 195; Wise 1980; Hounshell and Smith 1988, 301–4, 370.
80 Swanson oral history, 1996/1997, 57.

81 Goeddel oral history, 2001/2002, 24.

82 Goeddel, e-mail message to author, December 8, 2008.

83 Boyer oral history, 1994, 89.

84 Riggs oral history, 2005, 66.

85 For the press's influential role in science and technology coverage, see Nelkin 1995.

86 Clark 1978.

87 The university filed a patent application, "Synthesis of a Eucaryotic Protein by Microorganisms," on August 11, 1978, citing Rutter, Goodman, and John Baxter as inventors. A 1980 agreement, after the postdocs protested, gave the scientists conducting the insulin and growth hormone research a percentage share in any future patent royalties: John D. Baxter, Howard M. Goodman, and William J. Rutter to Roger Dietzel [sic], July 11, 1980, CU-6, box 60: Board of Patents Meeting, September 22, 1981, UC (System), Office of Technology Transfer, University of California.

88 Ullrich oral history, 1994/2003, 21.

89 Ibid., 29–30; *Regents of the University of California v. Genentech* (1990) C-90-2232, Seeburg testimony, April 20, 1999, 1028. Seeburg transcript courtesy of Robert Cook-Deegan.

90 "The Genen100: First 100 to Join Genentech," unpublished list courtesy of Dennis Kleid.

91 Heyneker oral history, 2002, 89.

92 "Invitation and Open House Invitation List," December 1, 1978, copies courtesy of Roberto Crea. Among those invited were the investment firms Kleiner & Perkins, Innoven Capital, the Hillman Company, the Mayfield Fund, International Industrial Development, Sofinnova, and Hambrecht & Quist; and the pharmaceutical companies Institut Mérieux, Hoffmann–La Roche, KabiGen, and Eli Lilly.

93 Middleton oral history, 2001, 17.

94 Perkins oral history, 2001, 7.

95 "Human Insulin: Seizing the Golden Plasmid," 195.

96 In 1978 Congress, as an inducement to investment, reduced the capital gains tax rate from 48 to 28 percent.

97 Robert A. Swanson to the Lubrizol Corporation, August 28, 1979; Shirley Liu Clayton to Fred Middleton, September 13, 1979 (confirming Genentech's receipt of Lubrizol's check for $10 million). Documents courtesy of Fred Middleton.

98 Dickson 1979, 494.

99 Kenney 1986, 96–97. For Hybritech's formation, see Jones 2005.

100 For Rutter's failed effort to form a nonprofit corporation at UCSF to commercialize biotechnology products, see Rutter oral history, 1992, 187–90. For Chiron's foundation and early development, see Penhoet oral history, 1998. Interviews with Rutter on Chiron history are in progress as of 2010. See Jong 2006 for a comparison of biochemistry department organization at Berkeley, Stanford, and UCSF and its relationship to biotech spin-off firms.

1 Louise Kehoe, "Genetic Engineering's Growing Commercial Importance," *New Scientist*, July 12, 1979, 86.

2 *Regents of the University of California v. Genentech* (1990) C-90-2232, Seeburg testimony, April 17, 1999, 937. I thank Robert Cook-Deegan for access to transcripts of Seeburg's testimony.

3 Seeburg et al. 1977.

4 For information on KabiVitrum, I am indebted to McKelvey 1996, chaps. 6, 7.

5 According to one source, a standard two-year treatment with natural growth hormone required extraction from fifty to a hundred human pituitaries. Elkington 1985, 83.

6 Boyer, e-mail message to author, December 26, 2009.

7 McKelvey 1996, 120.

8 Ibid., 121.

9 I thank Dennis Kleid for providing the contract date.

10 A Harvard business professor stated flatly: "The Genentech-Lilly agreement . . . created a template that influenced the evolution of the biopharmaceutical industry for the next thirty years." Pisano 2006, 86.

11 McKelvey 1996, 139.

12 Ibid., 143.

13 Seeburg et al. 1978.

14 Swanson oral history, 1996/1997, 55.

15 Crea oral history, 2002, 54.

16 *Regents of the University of California v. Genentech* (1990) C-90-2232, Seeburg testimony, April 17, 1999, 985.

17 Ibid., Seeburg testimony, April 20, 1999, 1023.

18 Robert A. Swanson and Dennis G. Kleid to Howard Goodman, November 9, 1978. 77-064-1, folder: "Goodman et al., Rutter et al., Deposit of Microorganisms," Office of Technology Transfer, University of California.

19 John D. Baxter and Howard M. Goodman to Peter Seeburg, February 22, 1979: 77-064-1, folder: "Goodman et al., Rutter et al., Deposit of Microorganisms," Office of Technology Transfer, University of California.

20 For details of the Seeburg-Goodman dispute, see Hall 1987, 281–83.

21 Ullrich oral history, 1994/2003, 32.

22 *Regents of the University of California v. Genentech* (1990) C-90-2232, Seeburg testimony, April 20, 1999, 1048.

23 William Stockton, "On the Brink of Altering Life," *New York Times Magazine*, February 17, 1980.

24 Bill Rutter made explicit the threat to the UC patent application in a note to an attorney representing the university. William J. Rutter to Lorance Greenlee, March 8, 1979, 77-064-1, folder: "Goodman et al., Rutter et al., Deposit of Microorganisms," Office of Technology Transfer, University of California.

25 Roger G. Ditzel, UC patent administrator, to Robert A. Swanson, April 18 1979, 77-064-1, folder: "Goodman et al., Rutter et al., Deposit of Microorganisms," Office of Technology Transfer, University of California.

26 *Regents of the University of California v. Genentech* (1990) C-90-2232, Seeburg testimony, April 20, 1999, 1035.

27 "Rival Claims Staked over Gene-Spliced Growth Hormone," *Medical World News*, August 6, 1979, 20–21.

28 Yansura oral history, 2001–2, 82.

29 Heyneker oral history, 2002, 101.

30 The litigation is summarized in chapter 6.

31 The biochemist Ronald Wetzel recalled hearing Goeddel's whoop. Noting that the scintillation counter was next to his lab, Wetzel added: "I'm pretty confident about this—it is a very strong memory." Wetzel's comment on a chapter 5 draft, June 10, 2009.

32 Heyneker oral history, 2002, 100.

33 Quoted in "Labs Tie for Human Growth Hormone," *Science News* 116 (July 17, 1979): 22. The formal report appeared in *Nature*; see Goeddel et al. 1979.

34 Heyneker oral history, 2002, 98; Ron Wetzel comment on a chapter 5 draft, June 10, 2009.

35 Kehoe 1979, 86.

36 Martial et al. 1979.

37 "First Successful Bacterial Production of Human Growth Hormone Announced," for release July 11, 1979, Corporate Communications, Genentech, Inc.

38 For the timing of the various announcements, see Saltus 1979.

39 "First Successful Bacterial Production of Human Growth Hormone Announced."

40 Gonzalez 1979, 701.

41 John Noble Wilford, "Human Growth Hormone Is Produced in Laboratory," *New York Times*, July 11, 1979, C1–C2.

42 Swanson quoted in, "First Successful Bacterial Production of Human Growth Hormone Announced."

43 "No More Dwarfs," *Economist*, July 14, 1979, 87–88.

44 For a critique of over-optimistic media coverage of genetic engineering, see Goodell 1980.

45 Swanson oral history, 1996/1997, 78.

46 According to Dennis Kleid's list, Genentech had 53 employees at the end of 1979. "The Genen100: First 100 to Join Genentech," unpublished list courtesy of Kleid.

47 Swanson, "Stanford Speech," unpublished manuscript, 1983, Swanson's office correspondence, K&E Management, San Mateo, CA.

48 U.S. Congress 1981, 67.

49 Swanson oral history, 1996/1997, 114.

50 "1979 Corporate Plan, Genentech, Inc.," Chief Financial Officer files, Genentech, Inc.

51 Robert A. Swanson, "Genentech—A Commercial Record of Achievement in Molecular Biology," unpublished presentation to E. F. Hutton, New York, NY, September 17, 1979, Swanson's office correspondence, K&E Management, San Mateo, CA.

52 Yansura oral history, 2001–2, 23.

53 For details of the scale-up of recombinant insulin and growth hormone, see Johnson 2003 and Cronin 1997. Genentech's scale-up process represents another instance of what John Lesch and others have termed the industrialization of pharmaceutical innovation. Lesch 2007.

54 Boyer, "The Business," unpublished presentation, "The Emergence of Biotechnology: DNA to Genentech" symposium, Chemical Heritage Foundation, Philadelphia, PA, June 13, 1997, 80–92.

55 Kleid oral history, 2001/2002, 97–99.

56 Quoted in "Gene-Splicing Factory Set to Produce Hormone," *Medical World News*, December 26, 1977, 17–18.

57 Swanson oral history, 1996/1997, 41.

58 "Insulin Research Raises Debate on DNA Guidelines," *New York Times*, June 29, 1979, A18.

59 Katherine Ellison, "Firm Pushes Ahead in Genetics," *Washington Post*, July 5, 1979, A9.

60 Ibid.

61 McKelvey 1996, 191.

62 Johnson 1983.

63 U.S. Congress 1981, 215, 218.

64 Gartland 1981.

65 Walgate 1980.

66 Yanchinski 1980.

67 For a summary of Lilly's human insulin trials, see Hall 1987, 299–301.

68 Genentech, Inc., Business Plan, December 1976, Chief Financial Officer files, Genentech, Inc.

69 "First Recombinant DNA Product Approved by the Food and Drug Administration," for release October 29, 1982, Corporate Communications, Genentech, Inc. According to one source, the first marketed recombinant product was a vaccine for pigs and calves made by the Dutch subsidiary Intervet. Yoxen 1983, 89.

70 Johnson 1983, 219–20.

71 "FDA-Approved Clinical Tests on Humans Begin Today with Human Growth Hormone Made by Recombinant DNA," for release January 12, 1981, Corporate Communications, Genentech, Inc.

72 Young oral history, 2004, 17–19.

73 "Tests Are Set for Synthetic Growth Hormone," *Chemical Week*, September 30, 1981, 41.

74 "Molecular Biology: The Industry Is for Real!—an Interview with Robert Swanson, President, Genentech, Inc.," *School of Management/ Massachusetts Institute of Technology* (magazine) (Winter 1982): 2–4.

75 Kleid oral history, 2001/2002, 88–89.

76 More than two dozen patients eventually died of Creutzfeldt-Jakob disease. Hall 1987, 248.

77 Harold M. Schmeck Jr. "U.S. Halts Distribution of a Growth Hormone as Precaution after 3 Deaths," *New York Times*, April 20, 1985, 9. Later research identified a prion as the causal agent.

78 "Genentech's Answer to Pituitary Derived Growth Hormone: Recombinant DNA Technology," for release April 22, 1985, Corporate Communications, Genentech, Inc.

79 Within a year, patients were paying as much as $10,000 to $15,000 a year for a drug that the government had earlier distributed at no cost. Hall 2006, 248–49.

80 "FDA Approves Genentech's Drug to Treat Children's Growth Disorder," for release October 18, 1985, Corporate Communications, Genentech, Inc.

81 Hamilton 1985, 108.

82 Bill Higgins to all employees, "Update on Upcoming Events/Employee Stock Option," internal correspondence, October 23, 1985, Corporate Communications, Genentech, Inc.

83 Hamilton 1986.

84 Kirk Raab to all employees, "HGH Project Team Organization," June 20, 1985, Corporate Communications, Genentech, Inc.

85 Glasbrenner 1986.

86 "Genentech Awarded Orphan Drug Status for Growth Hormone," for release December 17, 1985, Corporate Communications, Genentech, Inc.

87 Marcia Barinaga, "No Winners in Patent Shootout," *Science* 284 (June 1, 1999): 1752–53. For the beneficial and nefarious uses of human growth hormone—today marketed by a raft of companies in addition to Genentech—see Cohen and Cosgrove 2009. Genentech was not blame-free: in 1999 the company admitted that between 1985 and 1994 it had marketed Protropin for short but normal children, in violation of FDA rules against promoting drugs for unapproved uses. Cohen and Cosgrove 2009, 188–89.

88 "Genentech Product Development Summary," Genentech, Inc., 1979 Corporate Plan, Chief Financial Officer files, Genentech, Inc.

89 Kiley oral history, 2000/2001, 50.

90 Young oral history, 2004, 12–17.

91 Quoted in Benner 1981, 68.

92 Gower oral history, 2004, 48.

93 Yansura oral history, 2001–2, 83.

94 Swanson oral history, 1996/1997, 36.

95 Cape oral history, 2003, 62.

96 D'Andrade interview, 1998, 7–8.

97 Robert A. Swanson, "A Presentation to Henry F. Swift & Co.," April 27, 1983, Swanson's office correspondence, K&E Management, San Mateo, CA.

98 Fred A. Middleton to Bob Swanson, April 6, 1979, Chief Financial Officer files, Genentech, Inc.

99 Swanson, "Genentech—A Commercial Record of Achievement in Molecular Biology."

100 Goeddel, e-mail to author, April 14, 2009.

101 Packard 1995, 27. For earlier theories of research management in industrial laboratories, see Shapin 2008, chap. 6.

102 Swanson oral history, 1996/1997, 63.

103 Ronald Wetzel's comment on a chapter 5 draft, June 10, 2009.

104 Kleid's comments on a chapter 5 draft, April 23, 2009.

105 "Molecular Biology: The Industry Is for Real!—an Interview with Robert Swanson, President, Genentech, Inc.," 2–4.

106 Heyneker oral history, 2002, 93.

107 Yansura oral history, 2001–2, 28.

108 Crea oral history, 2002, 56.

109 Perkins 2007, 118.

110 Crea oral history, 2002, 48.

111 Ullrich oral history, 1994/2003, 43–44.

112 The six employees were Bob Swanson, administrative assistant Sharon Carlock, Brian Sheehan, Dave Goeddel, Dan Yansura, and Dennis Kleid. Kleid, e-mail message to author, March 18, 2009.

113 Kiley oral history, 2000/2001, 48.

114 For a description of the bizarre Ho-hos of the 1980s, see ibid., 49–50. The gatherings were later toned down, particularly after Hoffmann–La Roche's majority acquisition of Genentech in 1990.

115 Ibid., 48–49. For commentary on the phenomenon of "fun" in science, see Shapin 2008, 217–29.

116 For example, early Intel, reflecting president Robert Noyce's democratic outlook, offered employees a stock-option plan, almost identical cubicles, and no reserved parking places, regardless of status. Berlin 2005, 191.

117 At the end of 1979, only nine women of a total of fifty-three employees were employed at Genentech: "The Genen100: First 100 to Join Genentech," unpublished list courtesy of Dennis Kleid. Diane Pennica, who arrived at Genentech in May 1980, was the firm's first female scientist with a doctorate. She persevered in cloning and expressing tissue plasminogen activator (tPA), a treatment for blood clots. See Pennica oral history, 2003.

118 Gitschier 2004, 383.

119 Lasky oral history, 2003, 26.

CHAPTER 6

1 Robert E. Dallos, "Genetic Firm's Stock Starts Wall St. Frenzy," *Los Angeles Times*, October 15, 1980.

2 Marilyn Chase, "Industry Sees a Host of New Products Emerging from Its Growing Research on Gene Transplantation," *Wall Street Journal*, May 10, 1979.

3 Wade 1979.

4 Teitelman 1989, 26.

5 Cetus had alliances with Standard Oil of California and Indiana and with National Distillers and Chemical Corporation; Genentech's were

with Eli Lilly, KabiGen, Hoffmann–La Roche, Monsanto, and Institut Mérieux.

6 Sharon Begley with Pamela Abramson, "The DNA Industry," *Newsweek*, August 20, 1979, 53.

7 Statistic cited in Weiner 1982, 76.

8 Merrill Sheils with Gerald C. Lubenow, William J. Cook, Ronald Henkoff, Jeff B. Copeland, and Sylvester Monroe, "Innovation: Has America Lost Its Edge?" *Newsweek*, June 4, 1979.

9 For federal policy changes circa 1980, see Dickson and Noble 1981.

10 For reservations concerning the effect of the Bayh-Dole Act on stimulating commercialization, see Mowery et al. 2004 and Greenberg 2007, chap. 3.

11 "Statement of Stanley N. Cohen, M.D. Prepared for the Committee on Health, California State Assembly," April 25, 1977, Cohen correspondence, Department of Genetics, Stanford University. For a contemporary account of government regulation, risk assessment, and public perception of recombinant DNA research at the outset of the 1980s, see Weiner 1982.

12 See Wright 1994, 395–400, for RAC's concessions to industry. But Wright also maintained that industry and RAC members wished to keep some controls, arguing that abandoning the guidelines invited local intervention.

13 For these and other venture capital practices, see Wilson 1986.

14 Goeddel oral history, 2001/2002, 127.

15 Middleton oral history, 2001, 29.

16 Perkins oral history, 2001, 19–20.

17 Middleton oral history, 2001, 29.

18 Perkins oral history, 2001, 20. For a description of IPO strategy, see Southwick 2001, esp. chap. 1.

19 Glick 1982.

20 Interferon is technically a glycoprotein, a protein with carbohydrate attachments.

21 For the pre-genetic engineering history of interferon research, see Cantell 1998.

22 "The Big IF in Cancer," *Time*, March 31, 1980, 61.

23 Leukocyte and fibroblast interferon were subsequently renamed alpha and beta interferon, and immune interferon became known as gamma interferon. This book uses the older terms, current around 1980.

24 The confidential letter is not publicly available, but the later formal agreement references a letter of intent dated November 1, 1978. Dennis Kleid, e-mail message to author, July 24, 2009.

25 Robert A. Swanson, "Genentech—A Commercial Record of Achievement in Molecular Biology," unpublished presentation to E. F. Hutton, New York, NY, September 17, 1979. Swanson's office correspondence, K&E Management, San Mateo, CA.

26 Newspapers worldwide carried the story. See, for example, the front-page article by Harold Schmeck, "Natural Virus-Fighting Substance Is Reported Made by Gene Splicing," *New York Times*, January 17, 1980. For a

contemporary account of Biogen's interferon research, see Weissmann 1981.

27 Kiley oral history, 2000/2001, 86.
28 Gilbert, quoted in Schmeck, "Natural Virus-Fighting Substance Is Reported Made by Gene Splicing."
29 Bylinsky 1980, 145.
30 Taniguchi et al. 1979.
31 Wade 1980a.
32 Goeddel oral history, 2001/2002, 31.
33 Ibid., 30–31.
34 Ibid., 32–36; Goeddel, e-mail message to author, July 25, 2009.
35 "Genentech Announces Successful Production and Preliminary Biological Tests for Fibroblast and Leukocyte Interferon," for release June 4, 1980, Corporate Communications, Genentech, Inc. The research was published as D. V. Goeddel et al. 1980a and 1980b.
36 Harold M. Schmeck, "Exception to Ban on Genetic Material Is Backed," *New York Times*, June 11, 1980.
37 Wade 1980a.
38 In 1981 Genentech announced Goeddel's expression of "immune" interferon. Untitled news release, for release October 22, 1981, Corporate Communications, Genentech, Inc.
39 Pieters 2005, 172–73.
40 Bylinsky 1980, 45.
41 *BioEngineering News* 1, no. 1 (1980): 1.
42 Nelson Schneider, quoted in McAuliffe and McAuliffe 1981, 11.
43 As usual there were exceptions. Syntex, for example, staged an IPO without profits or products.
44 Perkins oral history, 2009, 52.
45 Perkins oral history, 2001, 12.
46 Quoted in Middleton oral history, 2001, 28.
47 Quoted in Perkins oral history, 2001, 12.
48 Middleton oral history, 2001, 29.
49 *Diamond v. Chakrabarty*, 447 U.S. 303 (1980). For a discussion of the *Chakrabarty* case and the history of patenting living things, see Kevles 1994.
50 Harold M. Schmeck, "U.S. to Process 100 Applications for Patents on Living Organisms," *New York Times*, June 18, 1980.
51 Hal Lancaster, "Rights to Life: Profits in Gene Splicing Bring the Tangled Issue of Ownership to Fore," *Wall Street Journal*, December 3, 1980.
52 Swanson oral history, 1996/1997, 46.
53 Kiley 1979, 3.
54 Ibid., 18.
55 Quoted in Jasanoff 1995, 144.
56 In 1988 the Patent Office issued a patent on an animal, the so-called Harvard mouse, engineered by Harvard and Genentech scientists for susceptibility to breast cancer. See ibid., chap. 7.

57 U.S. Congress 1981, 246.

58 *Amicus Curiae* Brief of the People's Business Commission. Decades later Rifkin continues to oppose patents on genes. See Rifkin 1998, chap. 2.

59 "Supreme Court Decision Will Spur Genetics Industry," June 16, 1980, Corporate Communications, Genentech, Inc.

60 Kiley oral history, 2000/2001, 39.

61 "The Hunt for Plays in Biotechnology," *Business Week*, July 28, 1980, 71.

62 The biotech companies Biogen, Hybritech, Collaborative Research, Molecular Genetics, Monoclonal Antibodies, Amgen, and Calgene formed in the 1978–80 period.

63 "The Hunt for Plays in Biotechnology," 71.

64 John D. Baxter, Howard M. Goodman, and William J. Rutter to Francis A. Sooy, Chancellor, UCSF, February 4, 1980, folder: "Genentech," Hormone Research Institute records, University of California, San Francisco.

65 Kiley oral history, 2000/2001, 22.

66 *Genentech, Inc., 1980 Annual Report*, 26. Copy courtesy of Dennis Kleid.

67 *Regents of the University of California v. Genentech* (1990) C-90-2232. Genentech, UC, and Eli Lilly, in various combinations, were participants in five lawsuits of the late 1980s and early 1990s, with many of the same people repeatedly testifying. The lawsuits were eventually consolidated, Genentech settled with Lilly, and the UC-Genentech claims were reduced to patent validity and infringement. Rimmer 2010.

68 Tom Abate, "Jury Gets Purloined DNA Case," *San Francisco Chronicle*, May 21, 1999.

69 *Regents of the University of California v. Genentech* (1990) C-90-2232, Seeburg testimony, April 20, 1999, 1087–96.

70 Hagmann 1999.

71 *Regents of the University of California v. Genentech* (1990) C-90-2232, Seeburg testimony, April 20, 1999, 1094.

72 Ibid., Goeddel testimony, May 4 and 5, 1999, 2411, 2547, 2786. Legal transcripts courtesy of Goeddel.

73 "UC-Genentech Trial," letter signed by Dennis Henner, David Goeddel, Herbert Heyneker, Keiichi Itakura, Daniel Yansura, Michael Ross, and Giuseppe Miozzari, *Science* 284 (May 28, 1999): 1465.

74 Seeburg's statement reads: "Not pHGH31 [the plasmid specified in Genentech's 1979 paper], but a functionally equivalent plasmid previously constructed by Shine and me at UCSF, was used as the source of cloned HGH complementary DNA in the construction of the expression vector." *Science* 284 (May 28, 1999): 1465. For the complexities in patent law that the lawsuit evoked, see Barinaga 1999a.

75 Barinaga 1999b. The five inventors on the UCSF growth hormone patent each received $17 million in the settlement.

76 Tom Abate, "Bitter Genentech Suit Finally Put to Rest," *San Francisco Chronicle*, November 20, 1999.

77 For the critical role of the outside attorneys, see Stewart 1980, chap. 3.

78 "The Hunt for Plays in Biotechnology," 71.

79 "Genentech Files Registration Statement with SEC for Its Initial Public Offering," for release August 19, 1980, Corporate Communications, Genentech, Inc.

80 N[icholas] W[ade], "Gene Splicer Goes Public," *Science* 209, no. 4461 (September 5, 1980): 1102.

81 Quoted in Timothy C. Gartner, "Stock's Amazing Day," *San Francisco Chronicle*, October 15, 1980.

82 Mike Johnson, "Behind Genentech's Decision to Go Public," *San Francisco Examiner*, August 20, 1980. The statement came from Genentech promotional literature, not directly from Swanson.

83 Middleton oral history, 2001, 30–31.

84 Stewart 1980, 137–38.

85 Ibid., 147.

86 Final Prospectus, Genentech, Inc., October 14, 1980, 5.

87 Ibid., 5–6.

88 The three remaining bottles are labeled Human Somatostatin, Human Thymosin Alpha-one, and Human Fibroblast Interferon. Final Prospectus, Genentech, Inc., October 14, 1980, color insert, fourth page, image D.

89 Middleton oral history, 2001, 30.

90 Ibid., 31.

91 "Frenetic Engineering," *Economist*, October 18, 1980, 108.

92 Quoted in Middleton oral history, 2001, 32.

93 Swanson oral history, 1996/1997, 101.

94 Genentech, Inc., *1981 Annual Report*, 26.

95 "New Genentech Issue Trades Wildly as Investors Seek Latest High-Flier," *Wall Street Journal*, October 15, 1980.

96 Perkins oral history, 2001, 10.

97 Charles J. Elia, "Genentech's Final Prospectus Shows Revenue Comes Primarily from Health-Care-Firm Jobs," *Wall Street Journal*, October 16, 1980.

98 Quoted in "New Genentech Issue Trades Wildly as Investors Seek Latest High-Flier."

99 Wade 1980c.

100 Susan Regan and Robin Stahl, "Herb Boyer: Hard Work Paid Off," *Greenburg Tribune-Review*, April 8, 1980.

101 Rothenberg 1984.

102 John March, "The Fascination of the New," *HBS* [Harvard Business School] *Bulletin*, October 1982, 55–62.

103 Judith Michaelson, "Genentech Soars: $300 in Stock Turns Buyer into Millionaire," *Los Angeles Times*, October 16, 1980.

104 Ullrich oral history, 1994/2003, 38.

105 David Dickson, "Genentech Makes Splash on Wall Street," *Nature* 287 (October 23, 1980): 669–70.

106 For Cohen's and Berg's reactions to the award, see Cohen oral history, 1995, 130–33; and Berg oral history, 1997, 122–26.

107 Watson, "Inexplicably neither Stanley Cohen nor Herbert Boyer has been so honored [with a Nobel Prize]." Watson 2002, 108.

108 "Lasker Awards Won by 4 Scientists from Bay Area," *San Francisco Examiner*, November 20, 1980, B4.

109 Yansura, e-mail message to author, September 28, 2009.

110 Stanford University News Service, advance for release August 3, 1981, S74-43, correspondence 1980–1982, Office of Technology Licensing, Stanford University. Two additional Cohen-Boyer patents were issued in the 1980s. When the three patents expired as a unit on December 2, 1997, they had earned over $250 million in royalties and license fees, at the time the highest income from any academic patent.

111 Hughes 2001.

112 "Frenetic Engineering," 108.

113 James Flanigan, "Apple Computer Looks Tasty but Its Market Needs Ripening," *Los Angeles Times*, October 19, 1980.

114 Farley statement in Johnson, "Behind Genentech's Decision to Go Public."

115 Perkins oral history, 2001, 10.

116 According to one source, the companies staging IPOs were Genetic Systems, Ribi Immunochem, Genome Therapeutics, Centocor, Bio-Technology General, California Biotechnology, Immunex, Amgen, Biogen, Chiron, and Immunomedics. Robbins-Roth 2000, 21.

117 "Man of the Year: Others Who Stood in the Spotlight: Herbert Boyer: Shaping the Future of Life," *Time* magazine, January 5, 1981, 38.

118 "Shaping Life in the Lab: The Boom in Genetic Engineering: Genentech's Herbert Boyer," *Time* magazine, March 9, 1981, cover.

119 A large literature exists on commercialism in the university. In addition to Greenberg 2007, see Bok 2003. For multiple perspectives on late modern life as an academic and/or industrial scientist, see Shapin 2008, chap. 7.

120 According to one source, 1980 alone saw the foundation of twenty-six biotechnology companies and more than forty in 1981. Then came an economic downturn with only twenty-two formed in 1983. Bud 1993, 193.

121 Yoxen 1983, 49.

122 Shapin defines the *scientific* entrepreneur "as one who is both a qualified scientist and, like all commercial entrepreneurs, a risk taker." Shapin, 2008, 210.

123 For the notion applied to biotechnology of "creative destruction" fueling innovation and progress as popularized by economist Joseph Schumpeter, see McKelvey 1996, 4–6; and Orsenigo 1989, 4–5.

EPILOGUE

1 Rathmann oral history, 2003, 16, 36.

2 Penhoet oral history, 1998, 110.

3 Quoted in U.S. Congress 1988, 128.

4 Quoted in Susan Brenner, "Genentech: Life under a Microscope," *Inc.*, May 1981, 62–68.

5 Kenney 1986, 4.

6 Gurin and Pfund 1980, 542.

Bibliography

Andreopoulos, Spyros. 1980. "Gene Cloning by Press Conference." *New England Journal of Medicine* 302 (March 27):13.

Barinaga, Marcia. 1999a. "No Winners in Patent Shootout." *Science* 284 (June 11): 1752–53.

———. 1999b. "Genentech, UC Settle Suit for $200 Million." *Science* 286 (November 26): 1655.

Benner, S. 1981. "Genentech: Life under the Microscope." *Inc.* (May): 62–68.

Berg, Paul, and Janet E. Mertz. 2010. "Personal Reflections on the Origins and Emergence of Recombinant DNA Technology." *Genetics* 184 (January): 9–17.

Berlin, Leslie. 2005. *The Man Behind the Microchip: Robert Noyce and the Invention of Silicon Valley.* Oxford: Oxford University Press.

Bliss, Michael. 1982. *The Discovery of Insulin.* Chicago: University of Chicago Press.

Bok, Derek. 2003. *Universities in the Marketplace: The Commercialization of Higher Education.* Princeton, NJ: Princeton University Press.

Boly, William. 1982. "The Gene Merchants." *California Magazine* 7:76–79.

Bud, Robert. 1993. *The Uses of Life: A History of Biotechnology.* Cambridge: Cambridge University Press.

Bylinsky, Gene. 1980. "DNA Can Build Companies, Too." *Fortune* (June 16): 144–46, 149, 152–53.

Cairns, John, Gunther S. Stent, and James D. Watson. 1966. *Phage and the Origins of Molecular Biology.* Cold Spring Harbor, NY: Cold Spring Harbor Laboratory of Quantitative Biology.

Cantell, K. 1998. *The Story of Interferon: The Ups and Downs in the Life of a Scientist.* Singapore: World Scientific.

Chang, Annie C. Y., and Stanley N. Cohen. 1974. "Genome Construction between Bacterial Species *In Vitro.*" *Proceedings of the National Academy of Sciences* 71 (4): 1030–34.

Clark, Matt, with Joseph Contreras. 1978. "Making Insulin." *Newsweek* (September 18): 3.

Cohen, Stanley N. 1975. "The Manipulation of Genes." *Scientific American* 233, no. 1 (July): 24–33.

——. 1982. "The Stanford Cloning Patent." In *From Genetic Engineering to Biotechnology*, edited by W. J. Whelan and Sandra Black, 213–16. Hoboken, NJ: John Wiley & Sons.

——. 1993. "Bacterial Plasmids: Their Extraordinary Contribution to Molecular Genetics." *Gene* 135 (1–2): 67–76.

Cohen, Stanley N., Annie C. Y. Chang, Herbert W. Boyer, and Robert B. Helling. 1973. "Construction of Biologically Functional Bacterial Plasmids *In Vitro*." *Proceedings of the National Academy of Sciences* 70, no. 11 (November): 3240–44.

Cohen, Stanley N., Annie C. Y. Chang, and Leslie Hsu. 1972. "Nonchromosomal Antibiotic Resistance in Bacteria: Genetic Transformation of *Escherichia coli* by R-Factor DNA." *Proceedings of the National Academy of Sciences* 69 (8):2110–14.

Cohen, Susan, and Christine Cosgrove. 2009. *Normal at Any Cost: Tall Girls, Short Boys, and the Medical Industry's Quest to Manipulate Height*. New York: Jeremy Tarcher/Penguin.

Crea, Roberto, Adam Kraszewski, Tadaaki Hirose, and Keiichi Itakura. 1978. "Chemical Synthesis of Genes for Human Insulin." *Proceedings of the National Academy of Sciences* 75, no. 12 (December): 5765–69.

Creager, Angela N. H. 2007. "Adaption or Selection?: Old Issues and New Stakes in the Postwar Debates over Bacterial Drug Resistance." *Studies in History and Philosophy of Science* 38, no. 1 (March): 159–90.

Cronin, Michael J. 1997. "Pioneering Recombinant Growth Hormone Manufacturing: Pounds Produced Per Mile of Height." Supplement 1, *Journal of Pediatrics* 131 (1): S5–S7.

Culliton, Barbara J. 1982. "Pajaro Dunes: The Search for Consensus." *Science* 216 (April 9): 155–56, 158.

Dickson, David. 1979. "Recombinant DNA Research: Private Actions Raise Public Eyebrows." *Nature* 278 (April 5): 494–95.

Dickson, David, and David Noble. 1981. "By Force of Reason: The Politics of Science and Technology Policy." *The Hidden Election: Politics and Economics in the 1980 Presidential Campaign*, edited by Thomas Ferguson and Joel Rogers, 260–312. New York: Pantheon.

Dutfield, Graham. 2003. *Intellectual Property Rights and the Life Science Industries: A Twentieth Century History*. Aldershot, UK: Ashgate.

Eisenberg, Rebecca S. 1987. "Proprietary Rights and the Norms of Science in Biotechnology Research." *Yale Law Journal* 97, no. 2 (December): 177–231.

——. 1996. "Public Research and Private Development: Patents and Technology Transfer in Government-Sponsored Research." *Virginia Law Review* 82, no. 8 (November): 1663–727.

Elkington, John. 1985. *The Gene Factory*. New York: Carroll & Graf.

Falkow, Stanley. 2001. "I'll Have Chopped Liver Please, or How I Learned to Love the Clone." *ASM News* 67: 555–9.

Fredrickson, Donald S. 2001. *The Recombinant DNA Controversy: A Memoir*. Washington, DC: ASM Press.

Gartland, William J. 1981. "Large-Scale Applications of Recombinant DNA Technology: Conduct under the National Institutes of Health (U.S.) Guidelines." In *Insulins, Growth Hormone, and Recombinant DNA Technology*, edited by John L. Gueriguian, 177–81. New York: Raven Press.

Genentech, Inc. 1981. Annual Report.

Gitschier, J. 2004. "Remembrances of Factor VIII. Part 1: The Race to the Gene." *Journal of Thrombosis and Haemostasis* 2 (3): 383–87.

Glasbrenner, Kimberly. 1986. "Technology Spurt Resolves Growth Hormone Problem, Ends Shortage." *Journal of the American Medical Association* 255, no. 5 (February 7): 581–87.

Glick, J. Leslie. 1981. "Genetic Engineering and Small Business Opportunities." In *Proceedings, 1981 Battelle Conference on Genetic Engineering*, vol. 1, edited by Melissa Keenberg, 13–18. Reston, VA: International Conference on Genetic Engineering.

Goeddel, David V., Herbert L. Heyneker, Toyohara Hozumi, Rene Arentzen, Keiichi Itakura, Daniel G. Yansura, Michael J. Ross, Giuseppe Miozzari, Roberto Crea, and Peter H. Seeburg. 1979a. "Direct Expression in *Escherichia coli* of a DNA Sequence Coding for Human Growth Hormone." *Nature* 28, no. 5732 (October 18): 544–48.

Goeddel, David V., Dennis G. Kleid, Francisco Bolivar, Herbert L. Heyneker, Daniel G. Yansura, Roberto Crea, Tadaaki Hirose, Adam Kraszewski, Keiichi Itakura, and Arthur D. Riggs. 1979b. "Expression in *Escherichia coli* of Chemically Synthesized Genes." *Proceedings of the National Academy of Sciences* USA 76, no. 1 (January): 106–10.

Goeddel, D. V., H. M. Shepard, E. Yelverton, D. Leung, R. Crea, A. Sloma, and S. Pestka. 1980a. "Synthesis of Human Fibroblast Interferon by *E. coli*." *Nucleic Acids Research* 8 (18): 4057–74.

Goeddel, D. V., E. Yelverton, A. Ullrich, H. L. Heyneker, G. Miozarri, W. Holmes, and P. H. Seeburg. 1980b. "Human Leukocyte Interferon Produced by *E. coli* is Biologically Active." *Nature* 287 (October 2): 411–16.

Gonzalez, Elizabeth Rasche. 1979. "Teams Vie in Synthetic Production of Human Growth Hormone." *Journal of the American Medical Association* 242, no. 8 (August 24/31): 701–2.

Goodell, Rae. 1980. "The Gene Craze." *Columbia Journalism Review* 19, no. 4 (November/December): 41–45.

Greenberg, Daniel S. 2007. *Science for Sale: The Perils, Rewards, and Delusions of Campus Capitalism*. Chicago: University of Chicago Press.

Greene, P. J., M. S. Poonian, A. L. Nussbaum, L. Tobias, D. E. Garfin, H. W. Boyer, and H. M. Goodman. 1975. "Restriction and Modification of a Self-Complementary Octanucleotide Containing the *Eco*RI Substrate." *Journal of Molecular Biology* 99, no. 2 (December 5): 237–61.

Gurin, Joel, and Nancy E. Pfund, 1980. "Genetic Engineering: Bonanza in the Bio Lab." *The Nation* (November 22): 542–48.

Hagmann., Michael. 1999. "Researcher Rebuked for 20-Year-Old Misdeed." *Science* 286, no. 5448 (December 17): 2249–50.

Hall, Stephen S. 1987. *Invisible Frontiers: The Race to Synthesize a Human Gene*. Redmond, WA: Tempus Books of Microsoft Press.

———. 2006. *Size Matters: How Height Affects the Health, Happiness, and Success of Boys—and the Men They Become*. Boston: Houghton Mifflin.

Hamilton, Joan. 1985. "Genentech Gets a Shot at the Big Time." *Business Week* (October 28): 108.

———. 1986. "Biotech's First Superstar: Genentech Is Becoming a Major-Leaguer—and Wall Street Loves It." *Business Week* (April 14): 68–72.

Hedgpeth, J., Howard M. Goodman, and Herbert W. Boyer. 1972. "The DNA Nucleotide Sequence Restricted by the R1 Endonuclease." *Proceedings of the National Academy of Sciences* 69, no. 11 (November): 3448–52.

Heyneker, Herbert L., John Shine, Howard M. Goodman, Herbert W. Boyer, John Rosenberg, Richard E. Dickerson, Saran A. Narang, Keiichi Itakura, Syr-yaung Lin, and Arthur D. Riggs. 1976. "Synthetic *lac* Operator DNA Is Functional *in vivo*." *Nature* 263, no. 5580 (October 28): 748–52.

Hotchkiss, Rollin D. 1965. "Portents for a Genetic Engineering." *Journal of Heredity* 56 (5): 197–202.

Hounshell, David A., and John Kenly Smith. 1988. *Science and Corporate Strategy: Du Pont R&D, 1902–1980*. Cambridge: Cambridge University Press.

Hughes, Sally Smith. 2001. "Making Dollars Out of DNA: The First Major Patent in Biotechnology and the Commercialization of Molecular Biology, 1974–1980." *Isis* 92:541–75.

Itakura, Keiichi, Tadaaki Hirose, Roberto Crea, Arthur D. Riggs, Herbert L. Heyneker, Francisco Bolivar, and Herbert W. Boyer. 1977. "Expression in *Escherichia coli* of a Chemically Synthesized Gene for the Hormone Somatostatin." *Science* 198 (December 9): 1056–63.

Jackson, David A., Robert H. Symons, and Paul Berg. 1972. "Biochemical Method for Inserting New Genetic Information into DNA of Simian Virus 40." *Proceedings of the National Academy of Sciences* 69:2904–9.

Jasanoff, Sheila. 1995. *Science at the Bar: Law, Science, and Technology in America*. Cambridge, MA: Harvard University Press.

Johnson, Irving S. 1983. "Human Insulin from Recombinant DNA Technology." *Science* 219 (February 11): 632–37.

———. 2003. "The Trials and Tribulations of Producing the First Genetically Engineered Drug." *Nature Reviews* 2 (September): 747–81.

Jones, Mark Peter. 2005. "Biotech's Perfect Climate: The Hybritech Story." 2 vols. PhD diss., University of California, San Diego.

Jong, Simcha. 2006. "How Organizational Structures in Science Shape Spin-off Firms: The Biochemistry Departments of Berkeley, Stanford, and UCSF and the Birth of the Biotech Industry." *Industrial and Corporate Change* 15, no. 2 (February 2): 251–83.

Kehoe, Louise. 1979. "Genetic Engineering's Growing Commercial Importance." *New Scientist* (July 12): 86.

Kenney, Martin. 1986. *Biotechnology: The University-Industry Complex*. New Haven, CT: Yale University Press.

Kevles, Daniel J. 1994. "Ananda Chakrabarty Wins a Patent." *Historical Studies in the Biological and Physical Sciences* 25:111–36.

Kiley, Thomas D. 1979. "Brief on Behalf of Genentech, Inc., Amicus Curiae." Supreme Court of the United States, October Term, no. 79–136.

Kleinman, Daniel Lee, and Steven P. Vallas. 2005. "Contradictions and Convergence: Universities and Industry in the Biotechnology Field." In *The New Political Sociology of Science: Institutions, Networks, and Power*, edited by Scott Frickel and Kelly Moore, 35–62. Madison: University of Wisconsin Press.

Kornberg, Arthur. 1995. *The Golden Helix: Inside Biotech Ventures*. Sausalito, CA: University Science Books.

Krimsky, Sheldon. 1985. *Genetic Alchemy: The Social History of the Recombinant DNA Controversy*. Cambridge, MA: MIT Press.

Lear, John. 1978. *Recombinant DNA: The Untold Story*. New York: Crown.

Lécuyer, Christophe. 2006. *Making Silicon Valley: Innovation and the Growth of High Tech, 1930–1970*. Cambridge, MA: MIT Press.

Lederberg, Joshua. 1975. "DNA Splicing: Will Fear Rob Us of Its Benefits?" *Prism*, November 33–37.

Lesch, John E. 2007. *The First Miracle Drugs: How the Sulfa Drugs Transformed Medicine*. Oxford: Oxford University Press.

Lewin, Roger. 1978a. "Profile of a Genetic Engineer." *New Scientist* (September 28): 924–26.

———. 1978b. "Modern Biology at the Industrial Threshold." *New Scientist* (October 5): 18–19.

Lowen, Rebecca S. 1997. *Creating the Cold War University: The Transformation of Stanford*. Berkeley: University of California Press.

Martial, Joseph, Robert A. Hallewell, John D. Baxter, and Howard M. Goodman. 1979. "Human Growth Hormone: Complementary DNA Cloning and Expression in Bacteria." *Science* 205 (August 10): 602–7.

Marx, Jean L. 1976. "Molecular Cloning: Powerful Tool for Studying Genes." *Science* 191 (March 19): 1160–62.

McAuliffe, Sharon, and Kathleen McAuliffe. 1981. *Life for Sale*. New York: Coward, McCann & Geoghegan.

McKelvey, Maureen. 1996. *Evolutionary Innovations: The Business of Biotechnology*. Oxford: Oxford University Press.

Mertz, Janet E., and Ronald W. Davis. 1972. "Cleavage of DNA by R1 Restriction Endonuclease Generates Cohesive Ends." *Proceedings of the National Academy of Sciences* 69:3370–74.

Morange, Michel. 1997. "The Transformation of Molecular Biology on Contact with Higher Organisms, 1960–1980." *History and Philosophy of the Life Sciences* 19:369–93.

Morrow, J. F., S. N. Cohen, A. C. Y. Chang, H. W. Boyer, H. M. Goodman, and R. B. Helling. 1974. "Replication and Transcription of Eukaryotic DNA in *Escherichia coli*." *Proceedings of the National Academy of Sciences* 71, no. 5 (May): 1743–47.

Mowery, David C., Richard R. Nelson, Bhaven Sampat, and Arvids Ziedonis. 2004. *Ivory Tower and Industrial Innovation: University-Industry Technology Transfer*

Before and After the Bayh-Dole Act in the United States. Stanford, CA: Stanford Business Books, 2004.

Nelkin, Dorothy. 1995. *Selling Science: How the Press Covers Science and Technology.* New York: W. H. Freeman.

Orsenigo, Luigi. 1989. *The Emergence of Biotechnology.* New York: St. Martin's.

Packard, David. 1995. *The HP Way: How Bill Hewlett and I Built Our Company.* New York: Harper Business.

Perkins, Thomas A. 2007. *Valley Boy: The Education of Tom Perkins.* New York: Gotham.

Pieters, Toine. 2005. *Interferon: The Science and Selling of a Miracle Drug.* London: Routledge.

Pisano, Gary P. 2006. *Science Business: The Promise, the Reality, and the Future of Biotech.* Boston: Harvard Business School Press.

Rabinow, Paul. 1996. *Making PCR: A Story of Biotechnology.* Chicago: University of Chicago Press.

Reich, Leonard S. 1985. *The Making of the American Industrial Research Laboratory: Science and Business at GE and Bell, 1876–1926.* Cambridge: Cambridge University Press.

Reimers, Niels. 1987. "Tiger by the Tail." *Chemtech* (August): 464–71.

Rifkin, Jeremy. 1977. "DNA: Have the Corporations Already Grabbed Control of New Life Forms?" *Mother Jones,* February–March, 23–26, 39.

———. 1998. *The Biotech Century: Harnessing the Gene and Remaking the World.* New York: Jeremy P. Tarcher/Putnam Inc.

Riggs, Arthur D. 1981. "Bacterial Production of Human Insulin." *Diabetes Care* 4, no.1 (January–February): 64–68.

Rimmer, Matthew. 2009. "Genentech and the Stolen Gene: Patent Law and Pioneer Inventions." *Bioscience Law Review* (December 7). http://pharmalicensing.com/public/articles/view/1070133214_3fc8efde16011/genentech-and-the-stolen-gene-patent-law-and-pioneer-inventions#f84 (accessed March 3, 2010).

Robbins-Roth, Cynthia. 2000. *From Alchemy to IPO: The Business of Biotechnology.* Cambridge, MA: Perseus.

Roberts, Richard J. 2005. "How Restriction Enzymes Became the Workhorses of Molecular Biology." *Proceedings of the National Academy of Sciences* 102, no. 17 (April 26): 5905–8.

Rothenberg, Randall. 1984. "Robert A. Swanson, Chief Genetic Officer." *Esquire,* December: 366–74.

Saltus, Richard. 1979. "Bay Scientists Find Way to Synthesize Hormones." *San Francisco Examiner,* July 11: 8.

Seeburg, Peter H., John Shine, Joseph A. Martial, John D. Baxter, and Howard M. Goodman. 1977. "Nucleotide Sequence and Amplification in Bacteria of Structural Gene for Rat Growth Hormone." *Nature* 270 (December 8): 486–94.

Seeburg, P. H., J. Shine, J. A. Martial, R. D. Ivarie, J. A. Morris, A. Ullrich, J. D. Baxter, and H. M. Goodman. 1978. "Synthesis of Growth Hormone by Bacteria." *Nature* 276 (December 21): 795–98.

Sgaramella, Vittorio. 1972. "Enzymatic Oligomerization of Bacteriophage P22 DNA and of Linear Simian Virus 40." *Proceedings of the National Academy of Sciences* 69, no. 11 (November): 3389–93.

Shapin, Steven. 2008. *The Scientific Life: A Moral History of a Late Modern Vocation.* Chicago: University of Chicago Press.

Sharpe, P. A., B. Sugden, and J. Sambrook. 1973. "Detection of Two Restriction Endonuclease Activities in *Hemophilus parainfluenzae* Using Analytical Agarose-Ethidium Bromide Electrophoresis." *Biochemistry* 12:3055–63.

Southwick, Karen. 2001. *The Kingmakers: Venture Capital and the Money behind the Net.* New York: John Wiley & Sons.

Stewart, James B. 1980. *The Partners: Inside America's Most Powerful Law Firms.* New York: Simon & Schuster.

Swann, John P. 1988. *Academic Scientists and the Pharmaceutical Industry: Cooperative Research in Twentieth-Century America.* Baltimore: Johns Hopkins University Press.

Sylvester, Edward J., and Lynn C. Klotz. 1983. *The Gene Age: Genetic Engineering and the Next Industrial Revolution.* New York: Charles Scribner's Sons.

Taniguichi, T., M. Sakai, Y. Fujii-Kuriyama, M. Muramatsu, S. Kobayashi, and T. Sudo. 1979. "Construction and Identification of a Bacterial Plasmid Containing the Human Fibroblast Interferon Gene Sequence." *Proceedings of the Japanese Academy* 55B: 461–69.

Teitelman, Robert. 1989. *Gene Dreams: Wall Street, Academia, and the Rise of Biotechnology.* New York: Basic Books.

U.S. Congress, Office of Technology Assessment. 1981. *Impacts of Applied Genetics: Micro-Organisms, Plants, and Animals.* Washington, DC: U.S. Government Printing Office.

———. 1988. *New Developments in Biotechnology: U.S. Investment in Biotechnology.* Washington, DC: U.S. Government Printing Office.

Vettel, Eric J. 2006. *Biotech: The Countercultural Origins of an Industry.* Philadelphia: University of Pennsylvania Press.

Wade, Nicholas. 1977. "Recombinant DNA: NIH Rules Broken in Insulin Gene Experiment." *Science* 197 (September 30): 1342–45.

———. 1979. "Recombinant DNA: Warming Up for Big Payoff." *Science* 206 (November 9): 663, 665.

———. 1980a. "Cloning Gold Rush Turns Basic Biology into Big Business." *Science* 208 (May 16): 688–89, 691–92.

———. 1980b. "New Horse May Lead Interferon Race." *Science* 208 (June 27): 1441.

———. 1980c. "Gene Splicing Company Wows Wall Street." *Science* 210 (October 31): 506–7.

Walgate, Robert. 1980. "How Safe Will Biobusiness Be?" *Nature* 283 (January 10): 126–27.

Watson James D., and John Tooze. 1981. *The DNA Story: A Documentary History of Gene Cloning.* San Francisco: W. H. Freeman.

Watson, James D., with Andrew Berry. 2003. *DNA: The Secret of Life.* New York: Knopf.

Weiner, Charles. 1982. "Relations of Science, Government and Industry: The Case of Recombinant DNA." In *Science, Technology, and the Issues of the Eighties: Policy Outlook*, edited by Albert H. Teich and Ray Thornton for the American

Association for the Advancement of Science, 71–97. Boulder, CO: Westview Press.

———. 1986. "Universities, Professors, and Patents: A Continuing Controversy." *Technology Review* (February/March): 33–43.

———. 1999. "Social Responsibility in Genetic Engineering: Historical Perspectives." *Gene Therapy and Ethics*, edited by Anders Nordgren, 51–64. Uppsala, Sweden: Acta Universitatis Upsaliensis.

Weissmann, Charles. 1981. "The Cloning of Interferon and Other Mistakes." In *Interferon 1981*, vol. 3, edited by Ion Gresser, 101–34. London: Academic Press, 1981.

Wilson, John W. 1986. *The New Venturers: Inside the High-Stakes World of Venture Capital*. Reading, MA: Addison-Wesley.

Wise, George. 1980. "A New Role for Professional Scientists in Industry: Industrial Research at General Electric, 1900–1916." *Technology and Culture* 21, no. 3 (July): 408–29.

Wright, Susan. 1986. "Recombinant DNA Technology and Its Social Transformation, 1972–1982." *Osiris* 2, 2nd series, 303–60.

———. 1994. *Molecular Politics: Developing American and British Regulatory Policy for Genetic Engineering, 1972–1982*. Chicago: University of Chicago Press.

Yamamoto, Keith. 1982. "Faculty Members as Corporate Officers: Does Cost Outweigh Benefit?" In *From Genetic Engineering to Biotechnology*, edited by William J. Whelan and Sandra Black, 195–201. New York: John Wiley & Sons.

Yanchinski, Stephanie. 1980. "More Freedom for US Genetic Engineers." *New Scientist* (June 26): 574.

Yi, Doogab. 2008. "Cancer, Viruses, and Mass Migration: Paul Berg's Venture into Eukaryotic Biology and the Advent of Recombinant DNA Research and Technology, 1967–1980." *Journal of the History of Biology* 41:589–636.

Yoxen, Edward. 1983. *The Gene Business: Who Should Control Biotechnology?* New York: Oxford University Press.

Oral History Bibliography

The oral histories for the Bancroft Library at the University of California, Berkeley, can be found in a searchable collection on bioscience and biotechnology at http://bancroft.berkeley.edu/ROHO/projects/biosci/.

Berg, Paul. 1978. Interview by Rae Goodell. Massachusetts Institute of Technology, Oral History Program, Oral History Collection on the Recombinant DNA Controversy, MC 100, box X. Massachusetts Institute of Technology, Institute Archives and Special Collections, Cambridge, Massachusetts.

———. 1997. *A Stanford Professor's Career in Biochemistry, Science Politics and the Biotechnology Industry.* Oral history conducted by Sally Smith Hughes, Regional Oral History Office, Bancroft Library, University of California, Berkeley, 2000.

Betlach, Mary C. 1994. *Early Cloning and Recombinant DNA Technology at Herbert W. Boyer's UCSF Laboratory in the 1970's.* An oral history conducted by Sally Smith Hughes, Regional Oral History Office, Bancroft Library, University of California, Berkeley, 2002.

Boyer, Herbert W. 1975. Interview by Rae Goodell. Massachusetts Institute of Technology, Oral History Program, Oral History Collection on the Recombinant DNA Controversy, MC 100, box X. Massachusetts Institute of Technology, Institute Archives and Special Collections, Cambridge, Massachusetts.

———.1994. *Recombinant DNA Research at UCSF and Commercial Applications at Genentech.* An oral history conducted by Sally Smith Hughes, Regional Oral History Office, Bancroft Library, University of California, Berkeley, 2001.

———. 2000. Interview by Sally Hughes and Arnold Thackray, draft. Chemical Heritage Foundation, Philadelphia.

———. 2009. "Wonderful Life: An Interview with Herb Boyer." Interview by Jane Gitschier. *Public Library of Science Genetics*5(9): e1000653.

Cape, Ronald E. 1978. Interview by Charles Weiner. Massachusetts Institute of Technology, Oral History Program, Oral History Collection on the

Recombinant DNA Controversy, MC 100, box X. Massachusetts Institute of Technology, Institute Archives and Special Collections, Cambridge, Massachusetts.

————. 2003. *Biotech Pioneer and Co-Founder of Cetus*. An oral history conducted by Sally Smith Hughes, Regional Oral History Office, Bancroft Library, University of California, Berkeley, 2006.

Cohen, Stanley N. 1995. *Science, Biotechnology, and Recombinant DNA: A Personal History*. An oral history conducted by Sally Smith Hughes, Regional Oral History Office, Bancroft Library, University of California, 2009.

Crea, Roberto. 2002. *DNA Chemistry at the Dawn of Commercial Biotechnology*. Oral history conducted by Sally Smith Hughes, Regional Oral History Office, Bancroft Library, University of California, Berkeley, 2004.

D'Andrade, Hugh A. 1998. Typescript of an interview conducted by Sally Smith Hughes and Leo Slater, Chemical Heritage Foundation, Philadelphia.

Falkow, Stanley. 1976. Interview by Charles Weiner. Massachusetts Institute of Technology, Oral History Program, Oral History Collection on the Recombinant DNA Controversy, MC 100, box X. Massachusetts Institute of Technology, Institute Archives and Special Collections, Cambridge, Massachusetts.

Gelfand, David. 1978. Interview by Charles Weiner. Massachusetts Institute of Technology, Oral History Program, Oral History Collection on the Recombinant DNA Controversy, MC 100, box X. Massachusetts Institute of Technology, Institute Archives and Special Collections, Cambridge, Massachusetts.

Glaser, Donald. 2003–4. *The Bubble Chamber, Bioengineering, Business Consulting, and Neurobiology*. Oral history conducted by Eric J. Vettel, Regional Oral History Office, Bancroft Library, University of California, Berkeley, 2006.

Goeddel, David V. 2001/2002. *Scientist at Genentech, CEO at Tularik*. Oral history conducted by Sally Smith Hughes, Regional Oral History Office, Bancroft Library, University of California, Berkeley, 2003.

Gower, James M. 2004. *Business Development and Marketing Strategy at Genentech, 1982–1992*. Oral history conducted by Sally Smith Hughes, Regional Oral History Office, Bancroft Library, University of California, Berkeley, 2006.

Heyneker, Herbert L. 2002. *Molecular Geneticist at UCSF and Genentech, Entrepreneur in Biotechnology*. Oral history conducted by Sally Smith Hughes, Regional Oral History Office, Bancroft Library, University of California, Berkeley, 2004.

Itakura, Keiichi. 2005. *DNA Synthesis at City of Hope for Genentech*. Oral history conducted by Sally Smith Hughes, Regional Oral History Office, Bancroft Library, University of California, Berkeley.

Johnson, Irving S. 2004. *Eli Lilly and the Rise of Biotechnology*. Oral history conducted by Sally Smith Hughes, Regional Oral History Office, Bancroft Library, University of California, Berkley, 2006.

Kiley, Thomas D. 2000/2001. *Genentech Legal Counsel and Vice President, 1976–1988, and Entrepreneur*. Oral history conducted by Sally Smith Hughes, Regional Oral History Office, Bancroft Library, University of California, Berkeley, 2002.

Kleid, Dennis G. 2001/2002. *Scientist and Patent Agent at Genentech*. Oral history conducted by Sally Smith Hughes, Regional Oral History Office, Bancroft Library, University of California, Berkeley, 2002.

Kornberg, Arthur. 1997. *Biochemistry at Stanford, Biotechnology at DNAX*. Oral history conducted by Sally Smith Hughes, Regional Oral History Office, Bancroft Library, University of California, Berkeley, 1998.

Lasky, Laurence. 2003. *Vaccine and Adhesion Molecule Research at Genentech*. Oral history conducted by Sally Smith Hughes, Regional Oral History Office, Bancroft Library, University of California, Berkeley, 2005.

Middleton, Fred A. 2001. *First Chief Financial Officer at Genentech, 1978–1984*. Oral history conducted by Glenn Bugos for the Regional Oral History Office, Bancroft Library, University of California, Berkeley, 2002.

Penhoet, Edward E. 1998. *Regional Characteristics of Biotechnology: Perspectives of Three Industry Insiders*. Oral history conducted by Sally Smith Hughes, Regional Oral History Office, Bancroft Library, University of California, Berkeley, 2001.

Pennica, Diane. 2003. *TPA and Other Contributions at Genentech*. Oral history conducted by Sally Smith Hughes, Regional Oral History Office, Bancroft Library, University of California, Berkeley, 2004.

Perkins, Thomas J. 2001. *Kleiner Perkins, Venture Capital, and the Chairmanship of Genentech, 1976–1999*. Oral history conducted by Glenn E. Bugos, Regional Oral History Office, Bancroft Library, University of California, Berkeley, 2002.

———. 2009, 2010. *Early Bay Area Venture Capitalists: Shaping the Economic and Industrial Landscape*. Oral history series conducted by Sally Smith Hughes, Regional Oral History Office, Bancroft Library, University of California, Berkeley, 2010. http://bancroft.berkeley.edu/ROHO/projects/vc/.

Rathmann, George B. 2003. *Chairman, CEO, and President of Amgen, 1980–1988*. Oral history conducted by Sally Smith Hughes, Regional Oral History Office, Bancroft Library, University of California, Berkeley, 2004.

Reimers, Niels. 1997. *Stanford's Office of Technology Licensing and the Cohen/Boyer Cloning Patents*. Oral history conducted by Sally Smith Hughes, Regional Oral History Office, Bancroft Library, University of California, 1998.

Riggs, Arthur. 2005. *City of Hope's Contributions to Early Genentech Research*. Oral history conducted by Sally Smith Hughes, Regional Oral History Office, Bancroft Library, University of California, Berkeley, 2006.

Rutter, William J. 1992. *The Department of Biochemistry and the Molecular Approach to Biomedicine at the University of California, San Francisco*. Oral history conducted by Sally Smith Hughes, Regional Oral History Office, Bancroft Library, University of California, Berkeley, 1998.

Scheller, Richard. 2001/2002. *Conducting Research in Academia, Directing Research at Genentech*. Oral history conducted by Sally Smith Hughes, Regional Oral History Office, Bancroft Library, University of California, Berkeley, 2002.

Swanson, Robert A. 1996/1997. *Co-founder, CEO, and Chairman of Genentech, Inc., 1976–1996*. Oral history conducted by Sally Smith Hughes, Regional Oral History Office, Bancroft Library, University of California, Berkeley, 2001.

Ullrich, Axel. 1994/2003. *Molecular Biologist at UCSF and Genentech.* Oral history conducted by Sally Smith Hughes Regional Oral History Office, Bancroft Library, University of California, Berkeley, 2006.

Yansura, Daniel G. 2001–2. *Senior Scientist at Genentech.* Oral history conducted by Sally Smith Hughes, Regional Oral History Office, Bancroft Library, University of California, Berkeley, 2002.

Young, William D. 2004. *Director of Manufacturing at Genentech.* Oral history conducted by Sally Smith Hughes, Regional Oral History Office, Bancroft Library, University of California, 2006.

Index